USING
AI
TO GROW YOUR
FRATERNITY
OR
SORORITY

By

Michael Ayalon and Ben Gold

CONTENTS

PRAISE FOR THIS BOOK

"Throughout the history of Greek-letter social groups, recruitment has always been a rush to secure the best and brightest potential new members. The thoughtful and purposeful use of AI can give fraternities and sororities a huge advantage in recruiting the strongest candidates. I wish the technologies and strategies detailed in this easy-to-follow book had been available during my time as an undergrad fraternity member. Mike and Ben have written the first MUST READ guide to recruitment in the 21st century."

—Dr. Ryan McKee,
Assistant Professor of Instruction in the College
of Public Health, Temple University

"This is a gold mine of resources and potential uses that are truly revolutionary. The use of AI to help drive understanding and put the skills of college students at the center of fraternity and sorority recruitment is exactly what we need. This book puts AI into an easy to understand guide that will be of great use to any chapter. I highly recommend sharing this book with anyone interested in supporting fraternity and sorority recruitment."

—Cory Bosco,
Director of Fraternity and Sorority Life at Bradley University

"As someone who has advised chapters in many different iterations and across councils, this book is an invaluable resource for recruitment. It provides prompts and examples for chapter recruitment, and even bonus material on job searches. AI is here to stay—we might as well use it to strengthen what we are doing and how we are doing it. I will be recommending this book to all the people I know who are interested in recruitment!"

—Dr. Kim Bullington,
Chief Departmental Advisor and Programs
Manager at Old Dominion University

"We live in a world where technology is proving its usefulness in the decision-making process. Mike and Ben offer a pragmatic approach in utilizing AI to streamline the recruitment process while offering ways to leverage skill development in students' careers. This easy-to-read guide will provide significant value to students and their chapters."

—Corbin Franklin,
PhD Candidate in Higher Education
Administration at Texas A&M University

PREFACE

by Michael Ayalon

CRITICAL THINKING CONSIDERATIONS

Here's the reason we should not release this book. There are concerns that if we widely use Artificial Intelligence (AI) within higher education, it could reduce your critical thinking skills as a student. It is critical for you as a leader within Fraternity and Sorority Life to develop sound decision-making skills and be able to process information on your own *without* the use of AI.

I recently spoke at Middle Tennessee State University to the Fraternity and Sorority community about leveraging your experience in college within Greek Life in order to get your dream job. It was a huge hit, the crowd cheered, we laughed, we learned, and we got extremely high scores from the student audience in our surveys that are conducted via cell phone during the presentation (data drives everything that I do). My son, Jacob, was in the audience, as he is a Data Science major at Middle Tennessee State University in his freshman year. I called him out on stage and I told the crowd that he was watching his dad on stage for the first time. Jacob was a celebrity of sorts, and students lined up to take selfies with him. Jacob thoroughly enjoyed it based on all the photos. Afterward, we went out to dinner with his best friend at college and had a great conversation about Artificial Intelligence over Thai food at a local restaurant.

Jacob's best friend had an older brother who had recently graduated from MTSU. He explained that his older brother was a graphic design major, and he was having trouble finding work in that field. I had to say a couple of things. First, I can hire anyone in the world today. One of my favorite books is *The World Is Flat* by Thomas Friedman. In it, the author explains that the playing field in terms of technology has leveled, and now you can find people to help your business from all over the world, and many times that's available at a fraction of the price. As a small business owner, if I wanted a graphic designer, I could go to fiverr.com and hire someone from anywhere in the world that has hundreds of positive reviews from past clients.

Second, jobs are shifting. We are at a moment in time where I can tell AI to create a video for me and the output would be 90 percent of what I want with no fee at all. It is likely that a graphic designer could get me 100 percent of what I want, but at what price? Is AI evolving at such a rapid pace that some jobs will be eliminated completely in the future? Yes. I would much rather give the students AI technology and have them play with it in real-life scenarios to figure out what AI does really well today and what the pitfalls are. In the case of our graphic designer, it is possible for him to learn

AI in order to speed up his creative process and take on more clients due to the efficiency it provides.

Third, your future employer will expect that you understand how to use AI to work more efficiently and get even better outcomes, so you might as well learn this new skill while you're in college. Regarding your critical thinking skills, I believe that the future is learning how to create the right prompt in tools like ChatGPT that deliver extraordinary results in the output. It does take critical thinking skills to develop the perfect prompt, and we will dive into that process in this book.

Let's talk about recruitment for a moment. We can call it rush or intake or recruitment depending on the council that you're in, but for simplicity's sake, we'll refer to it as recruitment in this book. Often during this formal recruitment period (usually the first two weeks of the semester), we don't have all of our brothers or sisters present at every recruitment event. Perhaps a few people are at work. Maybe a couple are doing homework. Perhaps a few more went home for the weekend to see their parents. How are we supposed to make critical decisions about who gets membership in our organization, when many of them haven't met all the potential new members (PNMs)? Also, there are so many new members that we can't possibly remember all of them or remember all of the conversations that we had with each of them. Or could we?

Today, we have access to social media such as Instagram or LinkedIn. Even if we just have a student's name, we can more than likely find one of their social media accounts. We should locate their social media accounts because Gen Z expects us to cater the experience to their needs/wants. You'll notice that they each have their own "brand" on social media in terms of the style and colors they use. Just based on who they follow on social media, that gives you a ton of information on where they are from (following local restaurants in Chicago, for example), which sports they like (following the Chicago Bulls and the Chicago Blackhawks), hobbies they enjoy (following multiple professional bodybuilders), and even political leanings (following multiple politicians from the Republican party). What if we could use all of that data for large numbers of students to create exactly the right text message for each PNM to get immediate positive responses?

DIVERSITY CONSIDERATIONS

There are some ethical considerations for collecting and using data for recruitment that we must discuss here. In the early 1900s, there were some organizations within Fraternity and Sorority Life that had exclusionary policies in terms of who they could invite into their organization. I am very happy to see these same organizations revise their bylaws and open up membership to diverse students over time. Gen Z has done a particularly good job with this, and I firmly believe they will continue to recruit more diverse students for their chapters.

I would argue that diversity is the most important thing you can have in a chapter because people from different places will have different lenses to solve the complex problems we face today. It is the diversity of our team at Greek University that allows us to make better decisions and faster decisions as a team. Our diverse team can point out all the things that I can't see, and I'm very grateful for that. You can't do that if your entire chapter came from the same high school or if they all belong to the football team. A great chapter can be seen through diversity in religion, race, country of origin, sexual orientation, or disability, and that will enable you to make better decisions, learn about new cultures, and grow as students.

Imagine this. What if we interviewed each PNM on Zoom for ten minutes and were able to transcribe all of that data? Could we give AI our fraternity and sorority values and find the top 20 potential new members based on their answers to our Zoom questions about our organization's values? Yes. What if we had the résumés of 100 PNMs—could we create a top 5 list that could fill in our hole in the Treasurer position because our current Treasurer is graduating in May? Yes. Could we do the same thing based on their LinkedIn profiles? Yes. What about written comments from our chapter members about each PNM— could that be used in our analysis of hundreds of PNM's? Absolutely. I'm expanding your mind when it comes to recruitment because there has been very little innovation in the area of Fraternity and Sorority recruitment over the last hundred years, and it's time to fix that in order to get the best new members in our organization. Unfortunately, because our Fraternities and Sororities are over one hundred years old, that means we are also generally slow to change or embrace new methods.

ALUMNI CONSIDERATIONS

I love the lifetime commitment that members of different councils show toward their organizations. Alumni regularly get together (wearing letters) to do service in their community, to network, and to mentor youth. In IFC or Panhellenic organizations, we have so much to learn in terms of how they are able to do that. What if AI could match up PNMs with the perfect alumnus that already works in their chosen field? Wouldn't the PNM be more likely to commit knowing that they already have a matched mentor in our organization before they even joined? AI can get us there today.

VALUES CONSIDERATIONS

I love Fraternity and Sorority Life. The pandemic was a big setback for many chapters across the country because we had to learn how to recruit on Zoom for a couple of years. Then, our expert recruiters took that institutional knowledge of how to recruit in person and graduated, leaving us with little experience or knowledge in actual recruitment. I want your organization to grow and be successful, and I see this book as an opportunity to help thousands of students rebuild their chapters. I'm committed to fixing some of the problems that we see all over college campuses such as hazing, alcohol/drug abuse, and sexual assaults through our amazing team of speakers and consultants at Greek University.

I was doing research recently for my dissertation on hazing prevention, and I came across a quote that made me pause, and it should make you pause too. According to Roberts and Rogers, "that the present condition of many fraternal organizations on today's college campuses resembles little of what their founders envisioned is an understatement. A beginning and core problem associated with the culture in fraternal organizations is that there are many members who did not join for the purposes espoused by the founders. This return to values movement, however, is dependent on the presence of a critical mass of dedicated and courageous undergraduates who are willing to take the responsibility to initiate change."[1]

Are you part of the dedicated and courageous undergraduates that are willing to take on that responsibility? If we are going to truly live our values within Fraternity and Sorority Life, protect our members and our guests at our house at all times, and continue to grow our organizations for the next hundred years, then we must realize that all of that depends on us choosing the right members. When we choose the wrong members that do not share our values, that's when we end up with people getting injured or worse. That's when we end up with apathy. That's when we end up unable to recruit, because we say we're about brotherhood, truth, justice, academics, respect for others, service, and leadership. When the PNMs see a member acting differently from the values we teach them, that's when we have misaligned expectations. They can get discouraged, ghost us, or drop out completely.

We are providing all of this to you and your organization in the hopes that you will use it to find the best new members

---

1   D. C. Roberts, D. C., and J. L. Rogers, J. L. (2003). "Transforming Fraternal Leadership". iIn Dennis. E. Gregory & Associates, *The Administration of Fraternal Organizations on North American Campuses: A Pattern for the New Millennium* (pp. 327 – 343). (Asheville, NC: College Administration Publications, 2003), 327–43.

on your campus, to ensure a good fit with your organization's values, and to grow your Fraternity/Sorority beyond your wildest dreams. We are also here to teach you how to fish. Artificial Intelligence can help us get the results we want and help us recommit to our founder's vision for the organization.

Navigating This Guide

This guide is the result of combined expertise from Michael Ayalon and Ben Gold, each eminent in their respective fields. Michael is a sought-after public speaker, author, and podcast host with extensive experience in Fraternity and Sorority Life. Ben brings to the table two decades of tech experience, with a recent focus on applying generative AI across various sectors. Together, they've curated this handbook to help fraternities and sororities effectively incorporate AI for recruitment and organizational management.

Chapter 1: Artificial Intelligence 101

Here, the foundations of AI are laid bare. You'll learn how to interact with AI tools like ChatGPT and Claude and understand key terminologies. You can opt to skip sections that cover familiar ground.

Chapter 2: Applying AI and Technology to Internal Processes

Dive into how AI can streamline your fraternity or sorority's internal workings. Topics include the use of AI in budget management, priority setting, and improving intra-organization communication.

Chapter 3: Outreach Activities for Recruitment

Discover how AI can revolutionize your recruitment strategies. Learn how to use AI-driven tools to generate content, objectively assess Potential New Members (PNMs), and prioritize candidates based on specific metrics.

Chapter 4: Bonus Section: Using AI to Boost Your Career

This chapter is your roadmap to a future where AI proficiency is a career necessity. It covers the use of AI for setting career goals, optimizing résumés, and effectively networking with alumni.

CHAPTER 1

ARTIFICIAL INTELLIGENCE 101

People Who Don't Use AI

People Who Use AI

People Who Master AI

NAVIGATING THE INTERSECTION OF ARTIFICIAL INTELLIGENCE AND FRATERNITY/SORORITY LIFE RECRUITMENT

When it comes to recruiting Potential New Members (PNMs) for fraternities and sororities, the focus has traditionally been on human interaction. We have relied on text messages, phone calls, social gatherings, and special events to forge meaningful relationships. After all, what's more human than selecting individuals who will join a lifelong brotherhood or sisterhood?

Yet, in today's increasingly digital world, Greek organizations, universities, and local chapters each possess unique characteristics that contribute to their recruitment strategies. This book aims to illustrate how technology, particularly Artificial Intelligence (AI), can enhance the recruitment process. By efficiently organizing data, processing vast information, and generating valuable insights, AI can empower leaders to make more informed decisions.

For the purpose of this book, we have conducted focus groups with fraternities and sororities ranging in size from seven to several hundred members. These groups vary in their approaches—from those who follow well-established data-sharing practices to those relying on gut instincts. We also explore organizations that utilize cutting-edge technological tools and those who employ minimal tech solutions.

At the core of this book is the aim to supplement and refine existing recruitment methods, all while preserving the irreplaceable value of human connection.

The goal of this book is to help fraternity and sorority leaders who are interested in growing their organization to embrace innovation and maintain a sense of exploration, especially with tools like ChatGPT and other state-of-the-art platforms.

AI and Career Development

One significant reason this book holds value for recruitment leaders is the transformative impact of generative AI across multiple roles and industries. The methodologies outlined herein are not exclusive to Fraternity and Sorority Life recruitment; they're actively being implemented by enterprises worldwide.

The utilization of AI technologies, whether through direct platforms like ChatGPT and Claude or via applications that incorporate these systems, is becoming a career imperative. As the job market continually evolves, mastering AI becomes essential for graduates who aim to maintain a competitive edge.

WHAT IS ARTIFICIAL INTELLIGENCE, AND HOW WILL IT HELP MY CHAPTER?

Artificial Intelligence (AI) is a branch of computer science concerned with creating and developing intelligent machines capable of performing tasks that would typically require human intelligence. These tasks include speech recognition, decision-making, visual perception, language translation, etc.

AI has been an important tool used by large corporations for some time. How many of the following do you recognize, and how do you think this technology could help fraternities/sororities?

> **Google Search Algorithms:** Google has been using AI in its searches for years, including algorithms like RankBrain, a machine learning-based algorithm, to provide more relevant search results.

> **Amazon's Recommendation System:** Amazon has been using AI to recommend products based on customer browsing and purchasing history. The AI learns from customer behavior to suggest items they are likely to be interested in.

> **Netflix's Personalization Algorithm:** Netflix uses AI to offer personalized recommendations to its subscribers. The AI learns from a user's viewing history and preferences and suggests TV shows and movies they might enjoy.

> **Spam Filtering:** Email service providers like Google's Gmail and Microsoft's Outlook have used AI for years to filter out spam emails and protect users from malicious threats.

> **Facebook's News Feed:** Facebook uses AI to decide what content to show in each user's news feed based on their preferences, interactions, and behavior on the platform.

> **Fraud Detection:** Banks and credit card companies have been using AI to detect unusual patterns of activity that may indicate fraud.

> ➤ **Voice Assistants:** Apple's Siri, Google's Assistant, and Amazon's Alexa use AI to understand and respond to voice commands.

GENERATIVE AI—THE NEW REVOLUTION

Generative AI is a subset of AI that takes things a step further. Instead of just learning from data and making decisions or predictions based on that data, generative AI models can create new data that is similar to, or a continuation of, the data it was trained on. This is done by understanding the underlying patterns or distribution of the input data and generating new data that follow these same patterns.

For example, generative AI is the technology behind creating realistic images, writing human-like text, composing music, or even creating new video game levels. It's also what powers the advanced language models like ChatGPT by OpenAI or Claude by Anthropic, which can generate human-like text based on the prompts given to it. While AI involves systems that can learn and make decisions, generative AI systems can also generate new, original content.

Quick Note: Language models are powerful tools but have limitations. They can often generate grammatically correct but semantically nonsensical text, as they lack a true understanding of the world and the text they generate. They can also replicate and amplify the biases in the data they were trained on.

Furthermore, language models can be unpredictable. A language model can produce vastly different results with the same prompt twice. This is because many models, especially large ones like GPT-4, inject randomness into their outputs to generate more diverse responses.

GLOSSARY OF IMPORTANT AI TERMS

AI (Artificial Intelligence): Technology designed to mimic human intelligence and behavior.

Generative AI: A type of AI that creates new, original content based on its learning, such as GPT.

ML (Machine Learning): A type of AI that enables a system to learn from data autonomously.

Natural Language Processing (NLP): A field of AI that enables machines to understand, interpret, and generate human language.

Large Language Model (LLM): A type of AI trained on vast amounts of text data to understand and generate human-like text based on the input it receives.

Input: Data you feed into the AI model is broken into two categories:

1. **Prompt**: The commands given to an AI model that guides its response.

2. **Contextual Data**: Information relevant to a specific situation that an AI model uses to form appropriate responses. Examples include résumés, job descriptions, and social media profiles.

Output: The response generated from your input.

OpenAI: An artificial intelligence research lab that created ChatGPT.

GPT (Generative Pre-Trained Transformer): A type of AI language model developed by OpenAI.

Token: In language models, a unit of text that the model reads. It could be as short as one character or as long as one word.

 ## CHATGPT VS. CLAUDE: CHOOSING THE RIGHT LANGUAGE MODEL

As you navigate this book, you'll encounter recommendations for deploying either ChatGPT or Claude. Both are potent Language Models (LLMs), each with their unique strengths and limitations.

Key Distinctions:

ChatGPT: Excelling in content creation and analysis of smaller data sets (under three pages of text), this model benefits from a more extensive training dataset.

Claude: Optimized for scrutinizing significantly larger data sets, Claude is your go-to choice for heavy-duty data analysis.

In subsequent chapters, we'll specify which model aligns best with the task at hand for optimal outcomes.

As of October 2023, **we recommend a subscription to ChatGPT 4.0 at $20 per month** for those seeking enhanced results (elaborated upon in later chapters). In contrast, Claude also offers a paid version; however, its primary benefit lies in giving paying customers priority access during high-demand periods.

GETTING STARTED WITH CHATGPT

(If you are already using ChatGPT, you can skip this section.)

Go to https://openai.com/blog/chatgpt, click on "sign up," and create an account.

You can use an existing Google, Microsoft,
Apple Account, or an email address.

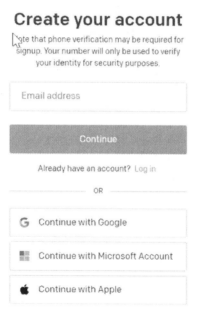

After successfully creating your account, you will be asked
to choose between ChatGPT, DALL-E, or API. DALL-E
is OpenAI's image creation application, while the API area
is how companies can integrate their websites and leverage
ChatGPT capabilities.

Do not create a group account because ChatGPT uses your cell phone as a unique identifier, therefore you would not be able to have a personal account.

After you click the ChatGPT box, you will notice a search bar with the text "Send a Message" at the bottom of the page. This is where all prompts and contextual data are added to get GPT output. Notice on the top that this is an example of a paid version. The program defaults to the GPT 3.5 version. For each chat, you need to manually click the GPT-4 button to use the GPT-4 version.

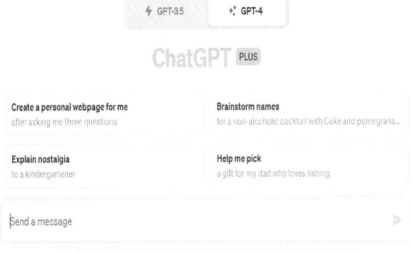

On the left of the screen, ChatGPT remembers every individual chat or conversation that you have had. The best practice is to create a new chat whenever you want to begin a new subject.

Advanced Tip: You can always return to older chats by clicking on the specific conversation. As you apply to different jobs, the best practice is to label each chat by opportunity. This way, it is much easier to leverage the contextual data you created (for example, from the job description and your résumé) when the opportunity moves to the interview phase.

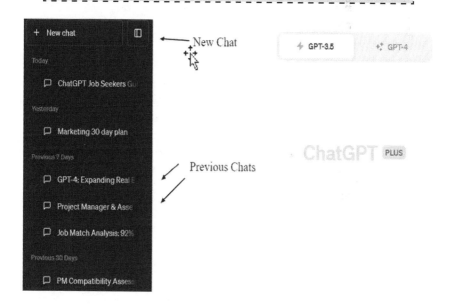

ChatGPT Paid vs. Free Versions

If you are brand new to ChatGPT, you can definitely take your time to familiarize yourself with how the AI functions.

You can test many of the prompts in this book and see the output. **Allocating a mere $20 per month for a ChatGPT 4.0 subscription should be an easy decision to incorporate into the Greek Life recruitment budget.**

Here are the differences:

- GPT 3.5 does offer faster output than 4.0 and is free. However, GPT 4 is trained on 10x the parameters as GPT 3.5.
- GPT 4.0 is smarter and more perceptive. This is crucial for creating new content or evaluating smaller sets of data.
- It has a larger memory. The GPT versions use a concept as tokens (1,000 tokens = about 750 words) to measure the amount of text it can evaluate and generate output. When you have a very long "conversation" with the AI, it will "forget" or no longer incorporate older data in its analysis.
- GPT 3.5 can retain 4,000 tokens, while 4.0 can retain 8,000 tokens. (For context, Claude can process over 100,000 tokens.)
- GPT 4.0 is multilingual, i.e., it can accept inputs in multiple languages.
- GPT 4.0, as of this writing, has a limit of 50 prompts every 3 hours.

▓ GETTING STARTED WITH CLAUDE

(If you are already using Claude, you can skip this section.)

Navigate to www.claude.ai

You can either use your email or Google account to login and then it will ask you a few more questions.

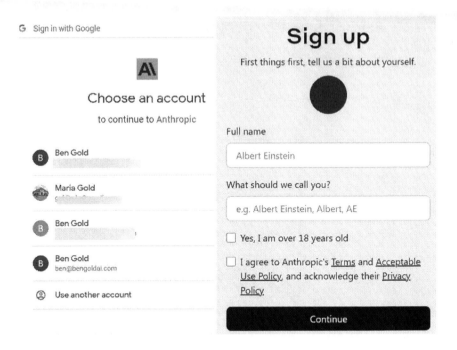

You will then be asked to provide a phone number for verification. Both ChatGPT and Claude use the phone number as the unique identifier. You can only have one account associated with a particular number.

Now you are ready to get started:

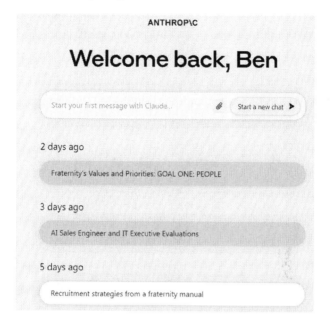

▩ INPUT #1—PROMPTS

A prompt is a command a user creates to generate a specific output from the AI model.

Prompts play a crucial role in determining the output of GPT models. The information contained in the prompt guides the model's text generation, influencing not just the topic but also the tone, style, and sometimes even the format of the generated text. People who have mastered this process are referred to as Prompt Engineers.

For instance, if you give the AI model the prompt, "Translate the following English text to French: 'Hello, how are you?,'" the model understands from the prompt that it needs to translate the text into French. Alternatively, if you input a more complex prompt like "Write a short, spooky story that starts in a haunted mansion," the model will generate a narrative fitting your specifications.

Understanding the Limitations of Prompts

While prompts are a powerful tool for guiding a GPT model's output, they have limitations. A prompt is only as good as the contextual data you supply. If a model doesn't have relevant data, it may struggle to generate a useful response, no matter how well-crafted the prompt.

For example, if you don't take the time to understand and document your career goals, skills, and achievements, the AI response will not yield insights that can give you actionable data.

IMPORTANT NOTE: Prompts can't guarantee accuracy. Even with a perfect prompt, GPT models might produce plausible-sounding but incorrect or nonsensical outputs. It's crucial to bear these limitations in mind when interpreting and using the output of GPT models.

A FOLLOW-UP BEST PRACTICE: NEVER POST OR SEND AI OUTPUT ELECTRONICALLY WITHOUT CAREFULLY READING THE TEXT AND VERIFYING IT TRULY EXPRESSES YOUR BEST INTERESTS.

Crafting Your First Prompt

> **NOTE:** If you have not yet set up your ChatGPT account, now is the time to put this book down and register. By doing the four examples below, you can get oriented on how generative AI works.

As a new user, the best way to understand how prompts work is to start simply.

→ **PROMPT:** *Give me a 5-day itinerary for New Orleans (or any other destination).*

You will notice that the AI will give you a generic response of a 5-day trip.

Let's add some more context:

→ **PROMPT:** *I am traveling with my college fraternity brothers.*

The AI will know what you just said refers to the previous prompt. There is no need to repeat the whole context. Notice how the itinerary has changed to focus on activities that are geared around young adults.

We can get more specific:

→ **PROMPT:** *3 days of seeing the best jazz clubs (or any other activity)*

The AI understands a group of college students is on a 5-day trip and wants 3 days to include jazz clubs.

Finally, let's dig deeper on day 2:

➜ **PROMPT:** *Expand on day 2 activities.*

As you begin interacting with ChatGPT, learning how to get big-picture responses and dig deeper into items you want to explore is important.

Congratulations! You have started on the journey of prompt mastery.

Tips for Crafting Effective Prompts

As you can see, prompts play an essential role in guiding a language model like ChatGPT to generate the desired output. Crafting effective prompts enhances the utility of these models to a great extent. Here are some tips for creating useful prompts:

- **Be Clear and Specific**: Precision and clarity are crucial when crafting prompts. The more specific your prompt, the more targeted the response will be. For instance, instead of "Write a story," you could state, "Write a short mystery story set in Victorian England involving a stolen artifact."
- **Guide the Tone and Style**: The language and tone of your prompt can help guide the model to generate text in a similar style. For instance, if you want a more formal output, use a formal tone in

your prompt. On the other hand, if you're looking for a creative, casual response, use a more casual tone.

- **Use Instructional Language:** Instructional prompts can be very effective, especially when you need the model to perform a specific task. For example, if you want a summary of a lengthy article, your prompt could be, "Summarize the following article in three concise sentences."

- **Set the Context:** If the task requires understanding a certain context, provide that context in the prompt. For example, if you're asking for a summary of a complex topic, you could start with a sentence or two about the topic to set the stage.

- **Experiment and Iterate:** It's important to remember prompt crafting can involve trial and error. Don't be afraid to experiment with different approaches and iterate based on the results you get.

- **Use Roles:** In a conversation with the AI model, you can use roles at the beginning to gently instruct the assistant about the behavior you want. For example, a message like, "You are an assistant that speaks like Shakespeare," can guide the model's tone throughout the conversation.

INPUT #2—CONTEXTUAL DATA

In AI and machine learning, contextual data refers to additional information that helps a system understand the circumstances or environment around a specific data point or an event. It's the auxiliary information that adds meaning to the primary data and allows a more nuanced interpretation of it.

Contextual data is critical for several reasons:

- Enhanced Understanding
- Improved Predictive Capabilities
- Reduced Ambiguity

Contextual data can come in many forms; however, in the context of Greek Life recruiting, there are two primary sets of data you'll use:

1. Data that defines values and goals that are important to your chapter:

- Information shared by national organizations that highlight their values and goals
- Information shared by local chapters regarding the top priorities of what they are looking for in PNMs

2. Data about PNMs that can be used to match against the goals and values of local and national chapters

- Data shared about PNMs during the initial recruitment phase
- Publicly available information on social media (Instagram/LinkedIn)
- Data collected that documents interactions including emails, zoom transcripts or notes from members about conversations with the PNM

CHAPTER 2

APPLYING AI AND TECHNOLOGY TO INTERNAL PROCESSES

THE LANDSCAPE OF FRATERNITY AND SORORITY RECRUITMENT

Greek Life, consisting of fraternities and sororities, is a cornerstone of many university experiences. They focus on brotherhood/sisterhood, leadership, community service, and personal development. Recruitment and marketing strategies are pivotal in attracting the right members and promoting the values of these organizations.

Current Methods of Recruiting

1. Formal Recruitment Events

Formal recruitment events are structured and organized activities designed to introduce potential new members to the fraternity or sorority, showcasing the organization's values, culture, and benefits.

2. Informal Recruitment Events, Usually Social Events

Informal recruitment events, often in the form of social events, provide a relaxed environment for prospects and members to interact, fostering organic connections and giving a glimpse into the social side of Greek Life.

3. Prospect Research

Prospect research involves gathering information about potential recruits to better understand their interests, backgrounds, and alignment with the organization's values, ensuring a good fit for both parties.

4. Networking with Brothers or Sisters and Alumni

Connecting with current members and alumni can provide potential recruits with deeper insights into the fraternity or sorority experience, highlighting long-term benefits and fostering a sense of community.

5. Branding

Branding emphasizes the unique identity, values, and culture of the fraternity or sorority, ensuring it stands out to potential recruits and resonates with the right audience.

6. Targeted Communication Strategies

Effective communication involves reaching out to PNMs through the right channels, with tailored messaging that speaks to their aspirations and interests, ensuring engagement and interest.

7. Fundraising

Highlighting successful fundraising initiatives showcases the organization's commitment to community service and philanthropy, attracting recruits who value social impact.

8. Projecting the Values Emphasized by a Chapter

Understanding and embodying the core values of a chapter ensures that both recruits and the organization are aligned in their goals, fostering a harmonious and purpose-driven community.

9. Mentorship Programs

Mentorship programs pair new recruits with seasoned members, facilitating smoother integration into the fraternity or sorority and fostering personal and professional growth.

10. Community Service Initiatives

Engaging in community service activities not only impacts the broader community positively but showcases the altruistic side of Greek Life, attracting service-minded individuals.

Current Recruitment Issues

Traditionally, Greek Life recruitment has involved face-to-face interactions, campus events, word-of-mouth, and maybe some print or digital advertising. Marketing might include promotional materials, social media campaigns, and events.

Not using technology to the fullest can lead to repetitive and inefficient efforts.

1. Recruitment Events Overlap

Organizations unknowingly schedule their recruitment events when other major events are happening on campus due to lack of centralized coordination tools, leading to a split audience and reduced attendance for both.

2. Time-Consuming Manual Data Entry

Members spend hours manually entering potential recruit data into spreadsheets, which could be prone to errors and inconsistencies.

3. Difficulty in Member Matching

Without a data-driven tool, an organization struggles to match recruits to suitable mentors, resulting in potential mismatches and friction.

4. Social Media Inconsistencies

Different chapter members posting inconsistently (or not at all) on social media due to a lack of streamlined communication, leading to mixed messaging and brand dilution.

5. Over/Under Ordering Promotional Materials

Relying on guesstimates, a sorority over-orders food or promotional materials for an event, leading to wasting precious resources.

6. Missed Communication

Using multiple communication channels without synchronization leads to some members being left out of important announcements or updates.

7. Manual Survey Analysis

After events, feedback surveys are manually processed (if done at all), delaying actionable insights and potential improvements for future events.

8. Event Planning Bottlenecks

Multiple members trying to coordinate events without a centralized tool results in double bookings, overlooked responsibilities, and missed opportunities.

9. Difficulty Tracking Engagement

Without automation, tracking alumni and member engagement in events, fundraising, or community service becomes cumbersome and inaccurate.

10. Struggle with Financial Transparency

Manual financial bookkeeping leads to challenges in budgeting, transparency, and potential financial discrepancies.

11. Limited Outreach

Using dated contact lists and manual email systems, outreach campaigns fail to connect with potential new members or alumni effectively.

12. Ineffective Alumni Relations

Without a proper database or communication tool, organizations struggle to keep alumni engaged, missing out on potential donations, mentorship opportunities, or networking.

13. Challenges in Diversity and Inclusion

A lack of data-driven insights might lead organizations to inadvertently ignore or overlook diverse groups during recruitment or event planning.

14. Event Feedback Lag

Feedback from events is gathered via paper forms or not gathered at all, causing delays in analyzing and implementing suggestions.

15. Inaccurate Member Records

Manually maintained member databases can become outdated quickly, leading to communication gaps and missed opportunities.

As we explore the chapters in this book, you will see how technology and AI solutions can help organizations become more efficient and effective and therefore reduce these problems.

The Importance of Data Gathering

Documenting the following three categories will allow chapters to make data-driven decisions on spending time and money on future efforts.

1. Define your goals7

At the national level, the organizational goals and priorities are spelled out on the website and in other documents. Use the document that best outlines priorities as part of the contextual data that is fed into the AI model.

Add a second document which spells out chapter level goals **and rank them as agreed by the entire chapter.** Be sure to revisit this every six months. This will save you lots of time when it's time to select members:

- Academics/scholarship
- Leadership abilities
- Community service history
- Integrity
- Respect for others
- Accountability
- Growth potential
- Chapter position needs (i.e., we have no Accounting majors to serve as Treasurer right now)
- Social/networking abilities
- Other

Be as specific as possible—is there a sport, religious affiliation, or academic major that fits the priorities of the local chapter?—write this down in the document.

2. Create a system for organizing PNM profiles

Some organizations use tools such as ChapterBuilder or OmegaFi which allows members to input data on PNMs in a systematic way. Others do not have a formal methodology.

If your chapter does not have a data aggregation tool, we suggest starting with a Google Doc or Google Sheet. It is important that a member of the recruitment team takes the time to organize the following items:

Demographic data shared at the early stage of recruitment:

- Name
- Phone Number
- Email
- Social Media account handles LinkedIn, X or Instagram (can be researched)
- Any other data that is shared publicly

Best practice suggestion: once a PNM shows interest in your particular chapter, have them fill out a simple questionnaire (Google Forms) that address these issues:

- Academic goals/GPA
- Leadership qualities
- Person priorities
- Other questions that give you insight

Document interactions including comments from other members.

- Events attended.
- Observations from chapter members

3. Document activities, promotional initiatives and expenses related to these events

For each event (informal/formal) sharing expense data and attendance data (number of attendees and specific attendee names), this will help you make better decisions in the future once the recruitment period is over. AI can help identify where the organization got the most ROI for identifying the events that led to the highest quality of interactions.

FINANCIAL MANAGEMENT AND BUDGETING

Budget Planning and Prioritization with a Focus on Recruitment:

1. Recruitment-Centric Data Analysis

To properly analyze past recruitment costs, examine financial records from previous recruitment cycles and categorize expenses into relevant segments related to events, advertising, tools, and overhead.

- **Events** - Break down and tally all costs associated with hosting recruitment events (such as venues, catering, transportation, activities, materials/ supplies, and any contracted staff). Itemize each event individually including both large-scale (such as campus-wide fairs) and small-scale (like meet-and-greets).

- **Advertising** - Compile data on any marketing and promotional expenses including printed materials like flyers and banners as well as digital initiatives like social media advertising, paid search, website costs, etc. Track both the medium/channel and the specific campaign/content.

- **Tools** - Look at costs of any software, platforms, or services used in the recruitment process for collecting applicant data, communicating with prospects, and facilitating the application process.

Isolating the granular costs specific to recruiting from general operating expenses allows Greek organizations to thoroughly analyze recruitment spending and performance.

2. Identify Patterns and Trends

Look for patterns in this recruitment-specific data. Did certain recruitment efforts yield more members? Were there particular events or advertising methods that had a higher return on investment (ROI)?

Example: If hosting recruitment seminars in the previous years led to a spike in membership compared to informal gatherings, then the trend suggests investing more in such seminars.

3. Budget Projection Based on Recruitment Outcomes

The AI can project recruitment budget requirements based on both costs and outcomes. If a recruitment method was cost-effective and successful in the past, it could be allocated a higher budget for the upcoming cycle.

4. Prioritize Based on Historical Success

Allocate budget segments based on the historical success of various recruitment strategies. Those methods and events that consistently yielded positive results should be prioritized.

ChatGPT 4.0 Example

(You will need the paid version as of this writing to be able to take a picture and have the AI analyze the data). Take your camera and snap a photo from either a piece of paper or the computer screen. Make sure the numbers are readable.

NOTE: If you have a more complicated spreadsheet that cannot display all of the data in a single screen, here are your options:

1. Take multiple pictures and explain in the prompt to evaluate them together.
2. Copy the cells and paste into ChatGPT with the explanation that you are copying a spreadsheet.
3. If the budget has more than 100 line items, and has too much data for ChatGPT to process, use Claude.

Item	Item Description	Expected Quantity	Price Per	Previous Cost	Expected Cost	Discounts	Actual Quantity	Actual Total
Rush Booklets	-	200	$1.40	$350.00	$280.00		55	$280.00
Rush Calendar	-	200	$0.15	$37.50	$30.00			$30.00
House Tours				$45.00	$0.00			$0.00
Rush Shirts	Shirts	75	$20.00	$1,500.00	$1,312.50	-		$1,407.66
Events								
A Term Event	Quad Games							
Brother Hunt	Campus, OK	1			$50.00			$50.00
Spikes 'n' Bites	Food, Spikeball nets, dice	60-80	$0.00	-	$200.00			$200.00
Fire and Ice	Meal	50-55 Monday	$30.00	$1,550.00	$1,550.00	Waiver		$1,250.00
Fire And Ice Transp.	Bus	1		$500.00	$500.00			$550.00
Bubble Soccer	Contract	1		$800.00	$750.00	Travel Fee Waived		$735.00
Car Smash (Car)	Car	1		$500.00	$750.00			$700.00
Car Smash Removal	Tow Truck/Uhaul	1		$100.00	$100.00	Waiver		$100.00
BBQ Olympics	Food & propane	50-60	-	$140.00	$150.00			$150.00
Barcade	-	55-65		$2,000.00	$1,000.00	Travel Waiver		$1,000.00
Barcade Transportation	Bus	1	-					$380.00
Casino Night	Prizes/Food	70-80	-	$650.00	$650.00			$650.00
Dodgeball	Gym	40-60	$0.00	$0.00	$0.00	Waiver	-	$0.00
SAE Search	Campus	50-60	$0.00	$0.00	$0.00		-	$0.00
Rush Video	Drone	0	$0.00	$0.00	-	-	-	-
Football Game Snacks	Food	50-70	-	$20.00	$100.00			$100.00
Bid Dinner	Food	45	-	$400.00	$400.00			$500.00
Total Cost				$8,592.50	$7,822.50			$8,082.66

➔ **PROMPT:** *Here is a picture of the fraternity budget for recruitment last year. What can you tell me about this data?*

ChatGPT RESPONSE:

Based on the provided image, which is a spreadsheet detailing a fraternity's recruitment budget for the previous year, here's what can be deduced:

Categories of Expenditures:

There are main categories like "Rush Booklets," "House Tours," "Rush Shirts," and "Events." Under "Events," there are various specific events like "Brother Hunt," "Spikes 'n' Bites," "Fire and Ice," etc.

Budget Columns:

The budget has different columns for:

- **Item Description:** Describes the specific item or nature of the expense.
- **Expected Quantity:** Estimated number of items or frequency of the expense.
- **Price Per:** Cost of each individual item or unit.
- **Previous Cost:** The cost that was incurred for the same item or event in a prior period.
- **Expected Cost:** The anticipated cost for the upcoming period.
- There also seems to be a column for ""Differences"" or ""Remarks"", although it"s partially visible.

Specific Insights:

- The most expensive expected costs are for ""Rush Shirts"," ""Fire and Ice Transportation"," and "Barcade Transportation".
- "Rush Shirts" are expected to cost $1,312.50.
- The expected cost for "Fire And Ice Transp." and "Barcade Transportation" is $1,550 and $1,000, respectively.
- There are several events like "Dodgeball", "SAE Search", "Rush Video", etc., where no costs were incurred previously, and none are expected for the upcoming period.
- The total cost previously was $8,592.50, and the expected total cost for the upcoming period is $7,822.50. This indicates a projected reduction in expenses.

Potential Savings and Overspends:

- Savings are anticipated in areas like "Rush Booklets", "Rush Calendar", "Rush Shirts", and "Bubble Soccer".
- However, there are areas where spending is expected to increase, such as "Brother Hunt" and "BBQ Olympics".

Nature of Events:

The events are diverse, ranging from games like "Dodgeball" and "Bubble Soccer" to food events like "BBQ Olympics" and "Bid Dinner".

Some events require transportation, indicating off-campus activities.

In conclusion, this fraternity has a well-structured recruitment budget. It details anticipated expenses and compares them to previous costs, providing a clear picture of where the fraternity expects to save money and where they anticipate additional expenses. The variety of events also suggests a mix of on-campus and off-campus activities to engage potential new members.

**

→ **PROMPT:** *What additional data do you need in order to give suggestions on improving the effectiveness of the budget?*

ChatGPT RESPONSE

To provide suggestions on improving the effectiveness of the budget, the following additional data would be useful:

Objectives and Goals:

Understanding the primary objectives for each event can help determine if the funds allocated align with those goals. For instance, is the goal of an event to maximize attendance, build a sense of community, or provide specific experiences?

Feedback from Previous Years:

• Were participants satisfied with the events?

- Which events had the highest attendance and engagement?
- Any complaints or areas of improvement pointed out?

Actual Cost vs. Budgeted Cost:

- Having a column for the actual amount spent can provide insights into budgetary accuracy and areas where financial planning can improve.

ROI (Return on Investment) for Each Event:

- Which events led to the highest recruitment numbers?
- Cost per new recruit can be a metric to determine the effectiveness of each event.

Detailed Breakdown of Costs:

- For events with significant expenses, a more detailed breakdown can help identify areas of potential savings. For example, what does the "Rush Shirts" cost include? Design, printing, material?

Alternative Quotes:

- Have multiple vendors been explored for services like transportation, printing, or catering? Comparing quotes can help in getting the best deals.

Historical Data:

- More years of budgetary data can provide trends, showing which costs are consistently rising or where savings have been achieved over time.

Membership Feedback:

- What do current members think about each event? Which events do they deem most valuable? Their feedback can provide insights into which events are core to the fraternity's culture and which can be modified or replaced.

Post-event Surveys:

- Collecting feedback immediately after each event can give insights into what went well and what can be improved.

External Factors:

- Are there external factors influencing the costs? For example, inflation, changes in venue rental rates, or shifts in student preferences.

Miscellaneous Costs:

- Are there hidden or unexpected costs associated with any events? Being aware can help in better planning.

Sponsorships and Partnerships:

- Information on any external funding or sponsorships can reduce costs. Can local businesses sponsor events or provide goods/services at discounted rates?

Cost-Benefit Analysis:

- For each event, compare the benefits received (like the number of recruits, brand building) to the costs incurred.

With this additional data, we can analyze the budget more deeply, identify inefficiencies, and make recommendations to enhance the effectiveness of the budget, ensuring that funds are used in the most impactful manner.

***

INTERNAL COMMUNICATION AND COLLABORATION

Whether you are using GroupMe, Slack, Discord, Teams, or any other internal communication tool, it is important that the data gathering strategy is communicated with the rest of the organization.

Suggestions:

- Establish clear guidelines on what data should be gathered on potential new members (PNMs) and how it should be recorded. This includes profile info, conversations, event interactions, social media activity, etc.

- Create specific channels/spaces for PNM data sharing in your internal communication platforms like Slack, Discord, etc. You can have separate channels for profile submission, event updates, conversation transcripts, etc.

- Appoint 1–2 members to oversee data collection and synthesis in a master spreadsheet/doc. They should regularly compile updates from members and share insights.

- Assign members specific PNMs to track if quantity is high. Ensure there are clear instructions on the format to submit their interaction updates to the master data doc.

- The recruitment committee should closely track event costs and attendance and submit to the master spreadsheet. Photos/videos can also help capture event success.

- Conduct training on how to effectively document PNM interactions and use any tracking software/tools. Guidance on insights to note about prospects.

- Schedule regular meetings to discuss PNM prospects using the collected data. Chance for members to share feedback that may be missed in written docs.

- Have a clear protocol for highlighting any red flags/issues on prospects based on data gathered. Helps mitigate risks.

- Leadership should frequently communicate the importance of thorough data gathering and how it helps the chapter recruit the best new members. Updates on success stories.

- Ultimately aim for a centralized knowledge base on prospects that the entire chapter can leverage to get to know PNMs better and nurture relationships.

This data collection strategy will allow the chapters to apply the techniques that are suggested in the following parts of the book.

RECRUITMENT MANAGEMENT

Keeping recruitment data in a centralized location is important for making the best decisions. This section will share best practices for chapters without a budget for software tools.

Data can be gathered in **Word (Google Docs) or Excel (Google Sheets)**. Given the need for chapter-wide participation, it is recommended to use a shared document in which multiple people can edit.

Here are the top 5 recruitment management software providers:

1. **ChapterBuilder** (https://chapterbuilder.com) - Provides recruiting and chapter management software designed specifically for fraternities and sororities.

2. **GreekTrack** (https://greektrack.com) - Specialized customer relationship management (CRM) platform to organize and track Greek recruitment.

3. **GiveGab** (https://givegab.com) - Fundraising software with Greek Life tools like rush registration, event management, and donor management.

4. **OmegaFi** (https://omegafi.com) - Offers Greek Life software tools including recruiting, finance, and chapter management functions.

5. **Greekbill** (https://www.greekbill.com) - Provides Greek life management software covering areas like membership, accounting, operations and recruitment.

Recruitment Management with Excel and Word (Google Sheets and Google Docs)

Initial PNM Lists (Before Contacting): Generally, there is a time when university students in the fall and spring signal a general interest in joining a Greek organization. This list is typically shared to all sororities and fraternities. The majority of the data collected is demographic information with some optional questions about hobbies and interests.

Initial Demographic data should be housed in a spreadsheet with all categories collected: Here are examples of categories that are collected:

- First Name:
- Last Name:
- Home-City:
- Home-State:
- Home-Zip:
- High School:
- High School City:
- High School GPA:
- Class Year"
- Major:Credit
- Hours Completed:
- University GPA:
- Service Involvement:
- Academic Involvement:
- Have you participated in a Greek Recruitment before?:
- If yes, where?:
- Date of recruitment:
- Did you sign a preference card?:
- Have you ever pledged a National Org?:
- Date Pledged:
- Date Pledge Expires:
- Hobbies and other interests:

Important Note: Customized Recruitment Approaches Across Universities

It's crucial to understand that each university employs its own unique methodology for initial recruitment into fraternities and sororities. As such, not every step outlined here may be applicable or relevant to your specific organization or campus.

For instance, some universities kick-start recruitment by engaging with first-year students, while others postpone this process until later in the academic year or even until sophomore year.

Regardless of these variations, the principle of centralized data collection remains universally important. Even if certain recruitment strategies discussed here don't directly apply to your situation, the foundational concept of effective data gathering should not be overlooked.

Adding Data to this list (before initial contact): On the shared spreadsheet, there should be additional columns where members can add links to social media profiles. The main three are LinkedIn, Instagram, and X (formerly Twitter). In the future, there could be other social media that could be used for data mining.

If members find other PNMs who are outside the list, this is a great place for them to document this data so that the team has a centralized database with general information.

Creating PNM Profiles in Word

While many chapters evaluate PNMs event by event, the best way to aggregate data is to add it specifically to the profile of a PNM. As we proceed in this book, you will see that the more data we aggregate about a PNM, the better the AI can help to stack rank or bring out positive elements of alignment and areas of concern.

Here is an example of how to use Word to create a profile:

Alberto Rodriguez
Dallas, Texas
Richardson High School
3.65 HS GPA, Class of 2023
Instagram: arodriguez90998
X: arodriguez767676
Basketball, Debate
Initial AI Analysis – 85% match:

- Strong academic performance in high school, including honors.
- Technology experience and skills in programming.
- Experience as a teacher intern shows ability to mentor.

Event 1: COMMENTS
Event 2: COMMENTS
Event 3: COMMENTS
Event 4: COMMENTS

The important change in behavior: make sure to update data based on individual profiles instead of events or other systems. This will allow the AI to make new analysis based on comments and other data that is added during the recruitment process.

CHAPTER 3

OUTREACH ACTIVITIES FOR RECRUITMENT

EXTERNAL COMMUNICATION: USING AI TO GENERATE CONTENT

Most ChatGPT users will be familiar with the following use cases in this chapter. It is important to know that using generative AI can help branding and messaging.

The following examples are designed to show how ChatGPT can be used to generate ideas, create promotional materials designed to connect with PNMs, and help efficiently organize recruiting strategies.

ChatGPT 4.0 is the best tool to give the most direct and crisp output. The following examples are showing results from a Jewish-oriented chapter. By supplying data about your chapter priorities, the content will be tailored toward your goals.

→ **PROMPT:** *Here are the founding principles of a national Jewish fraternity: [paste text]. I am the recruitment chairman of a local chapter and*

am looking for suggestions of recruitment events.
Give me 10 ideas.

ChatGPT RESPONSE:

1. **Leadership Panel**
   Invite local Jewish community leaders to
   talk about the importance of leadership
   and community involvement.

2. **Tzedakah Day**
   Organize a charity event where you and
   potential new members can volunteer at a local
   nonprofit, such as a food bank or hospital.

3. **Israel Awareness Night**
   Host a film screening or speaker series focused
   on Israeli culture, politics, or technology.

4. **Culinary Event**
   Introduce PNMs to traditional Jewish
   foods like Challah, Latkes, or Matzah ball
   soup, maybe even having a cook-off.

5. **Business and Networking Mixer**
   Invite alumni who have succeeded in business,
   government, or other sectors to share insights
   and offer networking opportunities.

6. **Spiritual Exploration**
   Host a Shabbat dinner or Havdalah service
   to explore the spiritual aspects of Jewish
   life in a non-pressurized environment.

7. **Sports Day**
   Hold a mini-Olympics with sports that
   are popular in Israel, like soccer or Krav
   Maga, to add a fun and competitive edge.

8. **Cultural Art Night**
   Feature Jewish music, dance, or art, possibly with participation from local artists or musicians.

9. **Intellectual Debate**
   Tackle a social or ethical issue that aligns with the fraternity's principles, perhaps inviting a guest moderator for a structured debate.

10. **Jewish Heritage Scavenger Hunt**
    Create a scavenger hunt around campus or your town, focusing on places of importance to Jewish history or culture.

****

IMPORTANT NOTE: Once you have shared the contextual data, there is no need to paste the same data again. In this case, the AI will know that follow up questions refer to the previous data.

➜ **PROMPT:** *Expand on the culinary event idea.*

ChatGPT RESPONSE:

The culinary event could be called "Flavors of Tradition: A Journey Through Jewish Cuisine." This event aims to connect potential new members (PNMs) and current fraternity members through the exploration and enjoyment of Jewish culinary traditions. It not only serves as an interactive icebreaker but also as a cultural education platform.

Agenda: Introduction (10 minutes)

Briefly introduce the significance of food in Jewish culture.

Cook-Off Segment (1 hour)

- Divide into teams: Each team will be tasked with preparing a traditional Jewish dish, such as Challah, Latkes, or Matzah ball soup.
- Equip each cooking station with the necessary ingredients and tools.
- Culinary History Snippets (intermittent)
- While teams are cooking, have mini-presentations or trivia questions about the history and significance of the dishes being prepared.

Judging (20 minutes)

Invite local chefs or community leaders to judge the dishes on taste, presentation, and how well teams explain the cultural significance of their dish.

Community Potluck (30 minutes)

After the judging, everyone samples the variety of dishes prepared.

Wrap-Up (10 minutes)

Announce winners and thank participants and judges.

Discuss how the event ties back to Alpha Epsilon Pi's values and mission.

Materials Needed:

- Cooking stations
- Ingredients for selected dishes
- Cooking utensils
- Plates, silverware, cups
- Projector for presentations
- Prizes for winners

Optional Features:

Recipe Cards: Hand out cards with the recipes of the dishes made, including historical facts.

****

➜ **PROMPT:** *Give me 5 ideas on how to promote this event on Instagram.*

ChatGPT RESPONSE:

1. Teaser Posts

Create a series of posts that showcase the various traditional Jewish dishes that will be featured during the event. Use high-quality images and brief descriptions to pique interest.

2. Countdown Stories

Utilize Instagram Stories to conduct a countdown leading up to the event. Each day could feature a different aspect of the event, such as the judges who will be attending, the ingredients, or the cooking stations.

3. Live Q&A Session

Host a live Q&A session on Instagram to discuss the event. You could invite a guest chef or a fraternity alumnus who is familiar with Jewish cuisine to help answer questions and build anticipation.

4. User-Generated Content (UGC)

Encourage current members to share their own favorite Jewish recipes or food memories on their personal Instagram accounts. Use a unique event hashtag to easily track and repost this content to the fraternity's official account.

5. Behind-the-Scenes (BTS) Snippets

Share short video clips or photos of the preparation process: selecting ingredients, setting up cooking stations, or creating event merchandise. This gives a sneak peek and builds excitement.

****

➔ **PROMPT:** *Give me 2 examples of emails I can send to PNMs to promote the event. The goal is to make them compelling and offer a hook.*

ChatGPT RESPONSE:

Email Example 1: Enhanced Initial Invitation

Subject: · Join the Culinary Event of the Year with (OUR CHAPTER)!

Dear [Potential New Member's Name],

Ever wondered how to master the art of making Shakshuka or perfect a Matzah Ball soup? Your culinary journey begins with (OUR CHAPTER)!

· Event: "A Taste of Tradition: A Culinary Journey Through Jewish Culture"

· Date: [Insert Date]

· Time: [Insert Time]

· Location: [Insert Location]

·· Special Guest: We're thrilled to have [Chef's Name], a renowned expert in Jewish cuisine, to guide us through the evening.

· Goodie Bags: Leave with a recipe book filled with traditional Jewish recipes and some ingredients to kick-start your culinary adventure.

Don't miss this mouthwatering opportunity! RSVP by [Insert RSVP Deadline] to secure your spot.

To flavors and friendships,

[Your Name]

Recruitment Chair

Email Example 2: Enhanced Reminder

Subject: Last Chance: Unforgettable Flavors Await at (OUR CHAPTER's) Culinary Event!

Dear [Potential New Member's Name],

Time's running out, and we don't want you to miss this culinary fiesta!

· Event: "A Taste of Tradition: A Culinary Journey Through Jewish Culture"

· Date: [Insert Date]

· Time: [Insert Time]

· Location: [Insert Location]

·· Live Demos: Watch [Chef's Name] whip up delectable dishes in real-time.

· Interactive Stations: Roll up your sleeves and try your hand at cooking with guidance from our fraternity brothers.

RSVP by [Insert RSVP Deadline] to make sure you're on the guest list for this gastronomic gala.

We can't wait to break bread with you,

[Your Name]

Recruitment Chair

****

➔ **PROMPT:** *Give me 5 examples of text messages that are conversation starters I can use to promote this event.*

ChatGPT RESPONSE:

Mystery Ingredient

"Hey [Name], ever guess the secret ingredient in a dish just by tasting it? You'll love what we're planning. ·"

Skill Sharing

"Hi [Name], ever wanted to up your cooking game? We're planning something fun that you won't want to miss!"

Culinary Adventure

"Hey [Name], how adventurous are you with food? We're cooking up something cool, literally!"

Taste Test

"Hi [Name], can you handle spice, or are you more of a sweet person? Got something fun brewing for both!"

Flavor Fiesta

"Hey [Name], ever mixed chocolate and chili? We're about to take your taste buds on a wild ride!"

****

EXTERNAL COMMUNICATION: USING AI TO GENERATE IMAGE CONTENT

Text to image creation has evolved rapidly and will continue to do so. There are many tools available that can help you generate images by using the same principles that are used to generate text output.

The following tools are some of the more prominent examples of image generation capabilities.

The advantage of these tools is that now you can create interactive and engaging photos that can be used for emails, websites and social media posts.

Our recommendation is to try out many tools and pick the one that best suits your needs.

AI Image Creation Tools – October 2023

1. **Midjourney** (https://www.midjourney.com/) is an independent research lab that offers a bot that generates stunning images from simple text prompts in seconds, which can be accessed through Discord.
2. **DALL-E 3** (https://openai.com/dall-e-3) is an AI system that can create realistic images and art from a description in natural language. It is owned by the same parent company of ChatGPT.
3. **Canva Magic Studio** (https://www.canva.com) - a new AI-powered design platform that offers generative AI tools made specifically for creative needs, use cases, and workflows.
4. **Jasper Art** (https://www.jasper.ai/art) is an AI art generator that turns your text prompt into amazing art and realistic AI-generated images in seconds.
5. **Adobe Firefly** (https://www.adobe.com/) is a family of creative generative AI models accessible as a standalone web application and through features powered by Firefly in Adobe's flagship apps.

General Suggestions for Using Text-to-Image Apps:

- **Rapid Evolution:** Be aware that the technology behind text-to-image apps is evolving swiftly. However, the AI still struggles with rendering human figures accurately. Issues like body part misrepresentation are common, especially in earlier versions.

- **Iteration Is Key:** Prepare to experiment with multiple iterations before settling on an image suitable for your social media posts.

- **Language Specifics:** Refrain from using terms like "fraternity" or "sorority" as these apps often botch the rendering of Greek letters. Use the word "college" as a workaround.

- **Content Guidelines:** If you aim to downplay the presence of alcohol in your generated images, make sure to explicitly include "no alcohol" in your text description.

- **Detail Matters:** Be as descriptive as possible in your text prompt to achieve a more accurate rendering. Use qualifiers such as "casually dressed," "smiling," "engaged," or "standing in front of a building" to fine-tune the image outcome.

- **Review and Revise:** Always double-check the generated image for inadvertent or sensitive content before publishing it online.

- **Diversity Inclusion:** Include the term "diverse" in your description to ensure the generated image represents a broad range of ethnic and racial groups.

Getting started with Midjourney

Go to www.midjourney.com

You can either click:

- The bottom left button "Get Started" which has detailed documentation on how to use Discord.

The bottom right button "Join the beta program"

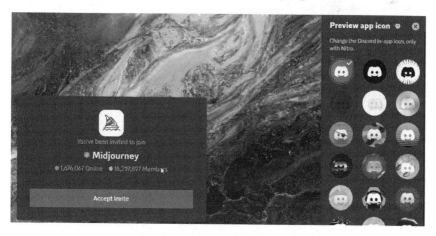

Accept the invite to join the Midjourney Discord server and you are in.

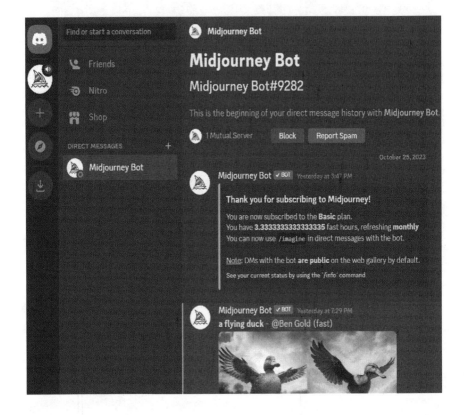

To create a new image, you need to type in:

/imagine (return) and then you will be given a prompt cue to start describing the image you want to create.

Midjourney will present 4 images and offer the following options once it is presented to you:

- U = Upscale (make the picture bigger without losing image quality)

- V = Create 4 Variations that are similar to that specific image
- Redo Icon allows you to have Midjourney generate 4 new images from the same prompt.

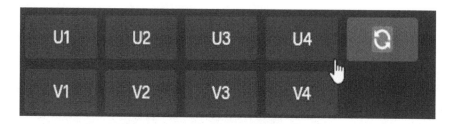

Midjourney also has commands that allow you to blend images and be more creative.

MIDJOURNEY VS DALLE - 3

- DALLE-3 is available at no cost via Bing search or through a premium ChatGPT subscription.
- Midjourney offers multiple subscription levels, starting at $10/month.
- As an early innovator in text-to-image technology, Midjourney delivers superior user experience for editing, refining, and amalgamating images.
- Certain experts lean towards DALLE-3 for its nuanced prompt interpretation.

Recommendation: Experiment with both platforms to determine which best facilitates your creative vision.

Actual Examples Midjourney vs DALLE-3

➔ **PROMPT:** *Create an image of a college party with diverse students having a great time dancing with no alcohol:*

DALLE-E 3 OUTPUT:

MIDJOURNEY OUTPUT:

➔ **PROMPT**: *Diverse college students playing mini golf*

DALLE-E 3 OUTPUT:

MIDJOURNEY OUTPUT:

➔ **PROMPT:** *Draw a picture of college students who are casually dressed at a bowling alley eating snacks with bowling balls in the background.*

DALLE-E 3 OUTPUT:

MIDJOURNEY OUTPUT:

****

USING AI FOR PNM DIRECT OUTREACH:PUBLIC DATA SOURCES

Creating profiles for PNMs can be difficult when there are in many cases, very few data points to go on. While every university has different policies on how data is collected and distributed, there are generally fall and spring recruitment seasons in which university students can show interest in joining a fraternity or sorority.

How can you leverage AI when all you have is a name, phone number, and city they come from? This usually is enough to locate someone's Instagram, X, or LinkedIn account.

Because the data needed to compile the following analysis is very large, **we suggest using Claude.**

In order to do this analysis, you will need to follow the instructions depending on which social media channel you use.

LinkedIn profiles and résumés are best options because they have aggregate data regarding a person's overall goals, activities, and achievements. Many high school students entering college do not have these available and hence it is suggested to review their Instagram or X profiles.

Using Instagram Profile Data

1. Locate the Instagram profile of the PNM
2. Click on the "following" text

Following ∨   Message   +

2,966 posts   466 followers   307 following

3.   Scroll down and copy as many examples as you can (this does take a minute or so); right click copy.

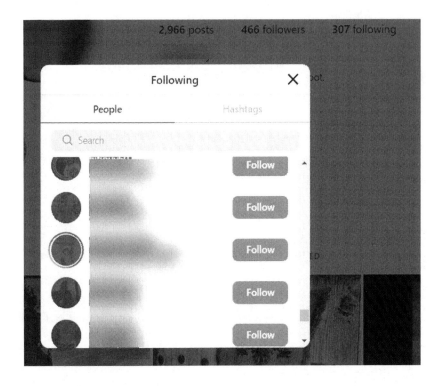

Using LinkedIn Profile Data

While not all college students use LinkedIn, if you are able to locate a PNM's profile, here are several ways to extract data and run a similar analysis.

First, download the PDF profile of the person you want to network with (located below their profile picture).

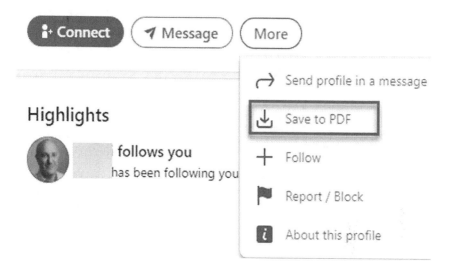

You can upload this document into Claude or copy and paste into ChatGPT, depending on the analysis you are making.

This PDF file has all professional and educational information. You can also gather additional insights by looking at the "Interests" section at the bottom on the page

Interests

Top Voices   Companies   Groups   Newsletters   Schools

Similar to Instagram, you can scroll through companies and groups (if they have any they are following) and add this data to the prompt.

Using X (formally known as Twitter) Profile Data

You can mine information from two areas: their post history and people they are following.

Post History: Click on their profile, scroll to their posts, and copy as far in the past as you can based on their actual posts.

When you copy the list of **people someone follows**, you click on the "following" text on the profile:

Then copy the list of followers and paste into ChatGPT or Claude.

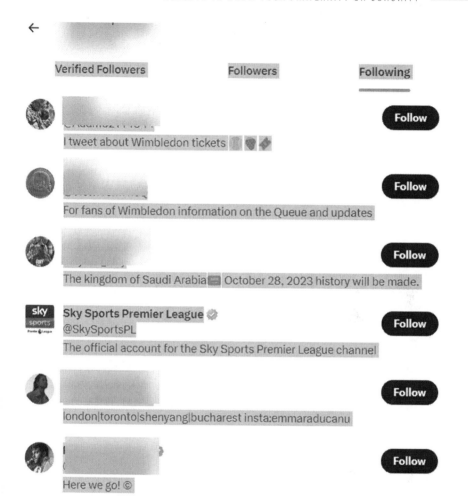

AI Output with Claude

→ **PROMPT:** *Here is a fraternity's values and priorities at a national level [paste text]. Here are people and groups that this person follows on Instagram. Based on those names, what can you tell me about this person and are there any insights on compatibility with our fraternity?*

> **NOTE:** If you use X, then add the following to the prompt: **"exclude ad text and recommendation suggestions in the evaluation."**

CLAUDE RESPONSE:

Based on the Instagram accounts this person follows, I can make a few observations:

- They seem to be interested in sports, particularly football, as they follow accounts like NFL, College Football Playoff, NFL Memes, and various football players. This suggests they are a sports fan.
- They likely have an interest in fitness and working out, following accounts like Athlete, HolStrength Ambassadors, and Spartan Elite Gains.
- They follow some accounts that suggest they have conservative political views, like Donald Trump Fans.
- They have an interest in cars/trucks based on following accounts like Hennessey Performance and Vanguard Motor Sales.
- They follow some Christian religious accounts like The Best Christian Memes, indicating they are likely Christian.
- They follow some entertainment/pop culture accounts like Barstool Sports, MrBeast, and Theo Von.
- They likely have ties to Texas based on following accounts like Texas Football Life, Texas MileSplit, and Dave Campbell's Texas Football.

Overall, this person seems to fit well with the interests and values of the fraternity, with overlaps in areas like sports, fitness, conservatism, Christianity, pop culture/entertainment, and an affiliation with Texas. Their Instagram follows don't reveal anything that would be incompatible with the fraternity. They seem to be a good fit based on common interests and values.

****

→ **PROMPT:** *I have his cell phone number. Give me 5 text messages that will likely make him interested in speaking with me about our fraternity. Focus on leadership, service, and academics and not on parties.*

CLAUDE RESPONSE:

1. Hey! I'm with [insert FRATERNITY] at your college. Our brotherhood is big on service and we raise tons for St. Jude Children's Hospital. Wanted to see if you'd be open to grabbing coffee and chatting about us.

2. What's up! Saw some of the accounts you follow and think you'd be a great fit for FRATERNITY. We're one of the largest fraternities and do cool stuff like relay marathons for charity. Let me know if you want to meet up and learn more!

3. What's up! I'm recruiting for FRATERNITY. Brotherhood and service are hugely important to us. We hold study hours and have scholarships too. Let me know if you want to learn more!

4. FRATERNITY here. I saw your profile and think you'd appreciate our values-based frat. Required study hours, plus we do Habitat for Humanity builds etc. Interested in discussing more?

5. Yo! Academics and service are key in FRATERNITY. We have one of the highest GPAs on campus and volunteer weekly. Thought you might appreciate our mission. Wanna meet up?

****

A different PNM analysis from the same fraternity data set:

→ **PROMPT:** *Here is a different prospect for the same fraternity. based on this person's Instagram profile, what is their compatibility to our fraternity as a % and explain why.*

CLAUDE RESPONSE:

Based on the Instagram accounts followed by this prospect, I would estimate their compatibility with the fraternity is around 65%. Here is my breakdown:

Positives:

• Follows many sports accounts, suggesting an athletic interest that aligns with the fraternity's active nature - 15% match

➔ **PROMPT:** *Here is one profile of a PN: [paste data]. What is her compatibility with the sorority?*

CLAUDE RESPONSE:

Based on the profile provided, here is my assessment of PNM's compatibility with the sorority:

Overall Compatibility: 85%

Values Alignment:

- PNM demonstrates a strong commitment to leadership development through her co-captain positions on varsity sports teams and role as student government representative. This aligns well with the value of embracing opportunities to develop leadership skills.
- Her involvement with her school paper and art club indicates an openness to gaining new perspectives and knowledge. This matches the value of being open to learning.
- As Community Outreach Director, she has shown initiative in serving her local community. This fulfills the value of contributing through service.
- Her various extracurricular activities and academic honors reflect drive towards achievement and excellence. This embodies the value of pursuing high ideals.

Code of Standards Alignment:

- PNM has demonstrated leadership, involvement in her community, and commitment to her academic studies. This fits with the standards around responsibility, contributing to one's community, and embracing leadership development opportunities.
- There is no indication of any conflicts with standards around proper conduct, adherence to rules, or presenting oneself properly.

Overall, PNM seems to strongly embrace many of the core values emphasized by the sorority. Her accomplishments and activities reflect an ambition to lead, serve others, and continue learning. With an 85% match, I would consider PNM to be a top-tier potential new member who would make a great addition to the sorority. Let me know if you need any clarification or have additional questions!

****

> **IMPORTANT NOTE:** By adding contextual data which can come from: emails, observations, documents such as a résumé, additional social media profile access, or responses to forms, these initial evaluations can be re-assessed by the AI and refined.

Let us take that same example from that initial 85% evaluation and add a new piece of information: **her résumé.**

➔ **PROMPT**: *Here is her résumé: [paste or upload résumé document].Reevaluate based on that additional data.*

CLAUDE RESPONSE:

After reviewing PNM's résumé, I would increase my assessment of her fit with the sorority to 90% compatibility. Here is my updated analysis:

Values Alignment:

- Her leadership roles as varsity sports captain and house representative demonstrate strong commitment to developing leadership abilities. She also co-led community outreach initiatives with She's the First, showing dedication to service.
- Her participation in multiple summer academic programs highlights an eagerness to learn new perspectives and skills, especially in art and design.
- Her camp counselor experience provides additional evidence of contributing through community service and leadership.

Code of Standards Alignment:

- PNM has clearly embraced opportunities to build leadership skills by taking on formal leadership roles in athletics, student organizations, and summer programs. This aligns well with standards on responsibility and leadership development.

- Her résumé paints a picture of an ambitious, high-achieving student who strives for excellence. This fits standards around academic achievement and pursuing high ideals.
- There are no conflicts with conduct standards evident from her résumé.

In summary, PNM's résumé provides further demonstration of her embodiment of key sorority values like leadership, achievement, and community service. The additional context boosts my assessment of her as a top-tier potential new member with a 90% compatibility match. Please let me know if you need any other details!

****

FRATERNITY EXAMPLE:

→ **PROMPT:** *We are looking at this particular PNM and have the profiles they follow on Instagram. Attached are the national HQ values and local chapter priorities: [paste both]. Based on these profiles, please share a % compatibility based on the values of the local and national chapter.*

CLAUDE RESPONSE:

Based on the Instagram profiles this potential new member is following, I estimate the compatibility with FRATERNITY's values to be around 65–70%.

Here is a brief analysis:

Positives:

- Following many FRATERNITY members and Greek Life profiles indicates interest in Greek Life and FRATERNITY specifically.
- Following campus groups like UNIVERSITY student associations and UNIVERSITY After Dark shows engagement with campus activities.
- Following a mix of male and female students is a good sign for building a positive brotherhood open to all.

Causes for concern:

- Following certain meme accounts could indicate potential for conduct issues if humor is insensitive.
- Lack of follows for service groups/social causes could signify the need to develop in service and philanthropy interests.

Overall, there seems to be decent initial compatibility but some areas FRATERNITY may need to nurture development further if this potential new member joins. Focusing on building knowledge of fraternity values, emphasizing service/charity, and mentoring positive personal conduct will help ensure a great fit. With proper education and role modeling from existing members, this person could likely become a highly compatible brother.

****

→ **PROMPT:** *Here is his LinkedIn profile: [paste profile]. Add this data and give a new assessment for compatibility.*

CLAUDE RESPONSE:

After reviewing this potential new member's LinkedIn profile, I would revise my compatibility estimate with FRATERNITY's values to around 80%. Here are some additional positive factors:

- Strong academic performance in high school, including honors, suggests a good fit with FRATERNITY's emphasis on scholarship.
- Technology experience and skills in programming align well with members who pursue tech-related majors.
- Experience as a teacher intern shows ability to mentor and assist others' development.
- Work experience in customer service roles indicates ability to positively represent an organization.
- Pursuing computer science and data science at UNIVERSITY fits with the school's strengths and many FRATERNITY members' majors.

Overall, the achievements and experiences highlighted on this LinkedIn profile demonstrate excellent compatibility with FRATERNITY's pillars of scholarship, leadership, service and character development. This PNM seems focused on personal growth and gaining technical skills, both of which align with the fraternity's vision to develop members to their full potential. I would feel very positively about this person's candidacy for membership in the FRATERNITY. Providing him mentorship and leadership opportunities could strengthen his development into an ideal brother.

****

OTHER EXAMPLES OF USING AI FOR PNM INTERACTIONS: GAINING DEEPER INSIGHT

AI can be used to prepare formal meetings and to get ideas on best engagement practices. Here are several examples of how AI can be used:

Building on the profile in the previous section, here are additional suggestions AI can do:

➜ **PROMPT:** *Based on the previous example, share 5 likely questions the PNM will ask and give me examples of recommended responses.*

CLAUDE RESPONSE:

Here are 5 likely questions this potential new member may ask along with example responses:

1. What kinds of community service and philanthropy work does FRATERNITY do?

- We have regular events like our annual Polar Plunge for Special Olympics and food drives. Brothers also volunteer through Habitat for Humanity, local animal shelters, tutoring programs, and more based on their interests.

2. How can I make sure my academics stay strong if I join?

- We have study hours, tutoring, and academic excellence programs to support our members' scholarship. Older brothers who have taken your courses before can give advice too.

3. What leadership opportunities are available in FRATERNITY?

- All members can take on leadership roles like heading a committee, organizing an event, becoming a pledge class officer, or joining the executive board. These opportunities help our brothers gain valuable experience.

4. How big of a time commitment is FRATERNITY?

- It varies week to week, but we try to balance activities with allowing time for schoolwork and other commitments. More events happen at the start of each semester usually.

5. What makes your chapter unique or special compared to others?

- Our chapter has a tight-knit brotherhood since UNIVERSITY has smaller Greek Life. Many brothers are also engineering/tech majors, so we support each other in tough classes. Our philanthropies and alumni network really distinguish us as well.

****

SUGGESTION: When ChatGPT or Claude touches on any idea of interest, it is important to learn to drill down and dig deeper to get more ideas as is in the next example:

→ **PROMPT:** *Give me 5 different answers to question #5.*

CLAUDE RESPONSE:

Here are 5 more examples of how I could respond to the question "What makes your chapter unique or special compared to others?":

1. We have one of the oldest FRATERNITY chapters, founded in 1896, so we have generations of tradition and alumni connections that make us stand out.

2. Our chapter house facilities were recently renovated and offer outstanding amenities like top-tier study rooms, dining hall, and recreation areas.

3. Many of our brothers are student athletes, so we have a culture of excelling both academically and in Division III sports like football, wrestling, and more.

4. Our chapter places a strong emphasis on personal development through leadership training, public speaking practice, career mentoring, and more.

5. We have a reputation on campus for hosting creative event themes, like our tropical luau philanthropy event or superhero brotherhood activities. Our social calendar is second to none!

****

Handling Objections

Generative AI tools are great for anticipating potential objections. By inputting detailed information about your fraternity or sorority, you can get well-crafted and personalized answers.

➔ **PROMPT:** *Help me handle the following objection with 5 examples: "I don't want to get hazed."*

CLAUDE RESPONSE:

Here are 5 ways to reassure a potential new member who says "I don't want to get hazed":

1. Our chapter has a strict no-hazing policy in line with FRATERNITY national guidelines and university rules. Your safety and well-being are our top priority.
2. Older brothers are carefully selected as new member educators and trained to lead constructive activities focused on teambuilding and education.
3. There are anonymous reporting options available if you ever feel uncomfortable with a situation during the new member process.

4. Hazing goes against our core values of friendship, service, and integrity. It has no place in our brotherhood.

5. I felt the same way when I was a potential new member. Our new member program brought me closer with my fellow new members and taught me about FRATERNITY's history in a meaningful way.

The key is stressing the educational nature of the new member program, the strict anti-hazing policies in place, assurance of appropriate oversight, and relatability to concerns based on your own positive pledging experience. This will help give PNMs confidence they will be treated with respect.

****

USING GENERATIVE AI FOR ADVICE ON PNM INTERACTIONS

There will be times when there are social media, email, or text interactions with PNMs and next steps need to be personalized. Generative AI tools help give creative ideas that can be quickly implemented.

Example: How to share conversations about a PNM that is no longer responding to texts:

→ **PROMPT:** *Here is a conversation between myself and a prospect for the fraternity: [paste fraternity goals and values]. Each paragraph return represents a change in who is writing. The* *first part is mine. What can you tell me about this prospect?*

- *Hey man! My name is Joe. I'm a part of FRATERNITY here at UNIVERSITY. I saw you on the GRADUATION YEAR class page and just wanted to reach out and ask if you have any questions about UNIVERSITY? If so, just give me a shout! Also, if you're interested in Greek Life I would be more than happy to give you any info you might want! Look forward to seeing you on campus bro!*
- *Definitely want to rush and info you got please let me know. Good to meet you.*
- *yeah bro! rush in general isn't really complicated here like it is at some of the other schools, basically you'll go to an event called convocation where you get to meet with all the different fraternities and see who you like, then you choose whose events you wanna go to for like 2 weeks and then you get your bids*
- *When will convocation week be?*
- *if i remember right it'll be at the end of the first week of school like that thursday or friday*
- *hey bro this is Joe from FRATERNITY at UNIVERSITY, was wondering when you're coming up for orientation? maybe we can link up with some of the guys and hangout*
- *Yeah that sounds great I'm going at the end of this month the 27th I think*
- *bet bro, add me on snap*

CLAUDE RESPONSE:

Based on this brief conversation, here are a few things I can infer about this prospect:

- He is explicitly interested in rushing a fraternity and Greek Life at UNIVERSITY. This indicates he is likely open to learning more about FRATERNITY.
- He seems eager to get information, asking follow up questions about the rush timeline. This suggests he is actively engaged and wants to be prepared.
- He is responsive and friendly in his messages, using language like "Good to meet you" and agreeing to meet up at orientation. This implies he is social and looking to make connections.
- He accepted the request to add on Snapchat, showing he is open to furthering the relationship beyond just texting.
- He is an incoming freshman for the class of 2027, so he can be a member for multiple years.

Overall, the prospect appears very receptive, socially inclined, and genuinely interested in Greek Life and FRATERNITY specifically. He seems like an ideal recruit based on the friendly rapport built already and willingness to continue the conversation on social media. Inviting him to hang out with some brothers is a great next step to get him engaged with the fraternity members. He displays promising signs as a potential new member.

****

→ **PROMPT:** *We played golf 4 weeks ago and had many conversations. He has ghosted me since then. Give me 4 ideas to re-engage with this prospect.*

CLAUDE RESPONSE:

Here are 4 ideas to try to re-engage with this ghosted prospect:

- Send a direct but lighthearted text like "Hey man, it's been a minute! We still on for meeting up during orientation? A few brothers are planning a campus tour and lunch if you wanna join." This re-establishes contact in a casual way.
- Comment or react to one of his social media posts to remind him you exist and are still interested in connecting. Just something simple to get back on his radar.
- Send him a text inviting him to an upcoming community service event or campus activity that FRATERNITY is participating in. Show him the fun/impactful things he's missing out on.
- Use a funny meme or inside joke referencing something you discussed before to break the ice and restart the conversation. Humor can disarm the situation. "That time we talked about [topic] made me think of this ..."

The key is to remain friendly and low-pressure while finding a natural way to remind him of the positive impressions he had of you and ORG originally. Meetings with potential new member classes coming up provide a solid context for outreach.

****

LEVERAGING AI FOR EFFICIENT PNM VOTING AND RANKING

The PNM selection process can be a high-stakes, time-sensitive affair, especially in fraternities and sororities that condense recruitment activities into a single week. Utilizing AI can significantly streamline and improve this process.

Situation 1 - No Membership Cap for Fraternity or Sorority

AI algorithms can rapidly pinpoint compatibility and red flags between PNMs and current members. This equips chapters with a strategic guide for targeted interactions and evaluations. Early identification of concerns allows for preemptive solutions, resulting in a smoother onboarding experience for new pledges.

Pre-recruitment data collection enhances AI's utility. Enriching PNM profiles with notes and observations allows for continuous refining of AI analysis. The outcome is more precise and actionable recommendations for PNM selection.

In essence, the more robust your data set, the more tailored and insightful the AI-driven suggestions become.

Situation 2 - Fraternity or Sorority with Membership Cap

When a fraternity or sorority has a membership cap, making the right choices becomes even more crucial. AI can assist by providing refined stack rankings based on data-driven metrics.

Green-Red-Yellow System:

Green: These are the "must-haves," either due to a strong emotional connection with members or a group consensus that the PNM fits perfectly with the chapter's ideals.

Red: These are the "must-not-haves," as they don't align with the chapter either academically, like a low GPA, or ethically.

Once you've filtered out both the Green and Red PNMs, what remains are the ones that require more nuanced evaluation—let's call these the "**Yellow**" candidates.

Data-Driven Evaluation for "Yellow" Candidates: This is where AI shines. By synthesizing a diverse range of data sources—from public information and PNM-submitted forms and emails to aggregated member comments—the AI algorithm provides a detailed stack ranking of these "Yellow" candidates. The result is a data-supported, unbiased assessment that helps in making more objective decisions.

FEEDBACK AND SURVEY TOOLS

A part of the data collection process can also be from post-event feedback. There are many companies that offer free and paid survey capabilities such as Survey Monkey, Typeform, and Jotform.

For the sake of simplicity and the price tag of Free, we recommend starting with Google Forms. Not every PNM will respond, however the ones that do will give you valuable feedback that you can both use in your analysis of the individual and in planning future events.

Here is an example of a post-event survey form:

Event Feedback Survey for [Fraternity/Sorority Name]

Thank you for attending [Event Name]. We value your feedback and would love to hear your thoughts to make our future events even better!

Personal Information

Name (Optional):

Are you a:

- Member
- PNM (Potential New Member)
- Guest

Event Experience

How did you hear about this event?

- Social Media
- Word of Mouth
- Flyer
- Email
- Other: [Text Field]

On a scale of 1 to 5, how would you rate the overall event?

- 1 (Poor)
- 2 (Fair)
- 3 (Good)
- 4 (Very Good)
- 5 (Excellent)

What did you enjoy the most about the event?

[Text Field]

Was there anything you didn't like about the event?

[Text Field]

Future Events

What type of events would you like to see in the future?

[Multiple Choice / Checkboxes]

- Social
- Philanthropic
- Educational
- Sports/Physical Activities
- Cultural
- Other: [Text Field]

Would you attend another event hosted by us?

- Yes
- Maybe
- No

Do you have any suggestions for improvement?

[Text Field]

Additional Comments

Any other comments or feedback?

[Text Field]

Thank you for taking the time to complete this survey. Your input is important to us.

****

CHAPTER 4

BONUS SECTION: USING AI TO BOOST YOUR CAREER

What career path is right for me? How to establish career objectives:

As you reach your Junior and Senior years, the question of what you will do after college becomes more and more pressing. Using AI can help you soul-search and understand your career options much quicker.

> SUGGESTION: I highly recommend plugging into your university career center and getting feedback on the work you do here. This stage is vital to the rest of the career journey, and your tuition is already paying for the people working in the university career center.

The other reason this stage is so critical is that most of the contextual data you will feed into ChatGPT will be based on your work here.

Formal self-assessment tools are available. If you have taken one recently, you can enter the results into ChatGPT for further analysis. Examples of these kinds of tests include:

- Myers-Briggs Type Indicator (MBTI)
- StrengthsFinder (Now called CliftonStrengths)
- Strong Interest Inventory
- DISC Assessment
- Self-Directed Search (SDS)
- CareerExplorer

If you want to begin this process without a formal skills assessment test, you can use a technique designed to answer behavioral interview questions. This is important so early in the process because your career journey is a story of who you are, what you're good at, and what you love to do. Concrete stories of when you accomplished important goals under pressure are part of that story.

I will share one of these techniques known as **STAR**:

- **S:** Describe the Situation.
- **T:** What Tasks were involved?
- **A:** What Action did you take?
- **R:** What was the Result?

SUGGESTION: Write down as many stories as you have in your university experience including projects you worked on at your fraternity or sorority that show leadership capabilities. In addition, if you have had the chance to gain professional experience, make sure to include elements that show when you have achieved results or gained satisfaction.

Here is an example from Ben's previous role as a Solution Consultant:

Situation: *The organization had a global sales force without any demo collateral that showed the company's global capabilities. Sales reps were using live client data to show business value.*

Task: *The challenge was to create a single global demo entity that highlighted the linguistic and localization capabilities.*

Action: *After evaluating the disconnected and illogical demo environments, I created a single streamlined organization that reflected situations the larger global clients faced. This included uploading documentation that resonated with UK, French, German, Australian, and Mexican audiences.*

Results: *The new GlobalCo demo environment was adopted by the whole organization and used to expand its global reach.*

I am sharing a long response in this example to show how ChatGPT can take a small amount of information and provide a very insightful and detailed response.

➔ **PROMPT:** *What conclusions can you draw about my skill set, goals, and professional roles I should pursue based on this STAR analysis?*

ChatGPT RESPONSE:

Skills:

- Project Management and Organizational Skills: Your ability to evaluate and restructure the demo environments signifies your strong project management and organizational capabilities.
- Cross-cultural Understanding and Adaptation: The successful creation of documents for diverse international audiences indicates a strong understanding of cross-cultural communication and the ability to adapt materials to fit different cultural contexts.
- Technical Skills: Your involvement in creating a demo environment suggests that you have good technical skills, possibly in software development, data management, or related fields.
- Strategic Thinking and Problem Solving: You were able to identify a major issue within the organization and devise a comprehensive solution that addresses not only the immediate problem, but also potential future issues, demonstrating strategic thinking and problem-solving capabilities.

Goals:

- Process Optimization: Your aim appears to be improving and streamlining processes within the organization, suggesting an interest in continuous improvement and efficiency.
- Global Engagement: You seem to have a goal of enhancing the global reach and impact of organizations, which can involve working closely with international teams, understanding global markets, and creating solutions that have wide appeal.

Professional Roles:

- Product Manager or Project Manager: Your demonstrated skills and goals align well with roles such as a Product Manager or Project Manager, where you would be in charge of overseeing the development and execution of projects, often with a global reach.
- Global Strategy Consultant: With your experience in global outreach and understanding of multicultural aspects, a role in global strategy consulting could be a good fit. You would assist organizations in understanding and entering new international markets.
- Sales Enablement Manager: Given your experience in creating demo environments and materials for sales teams, a role in Sales Enablement would be well-aligned. This could involve developing tools,

content, and training to improve the productivity of sales teams.

****

As you get more comfortable with prompting, you will learn that it is okay to go "off script." Some AI answers might not resonate with you. Others might be intriguing. AI will give you better output by sharing more data with the model. As you find interesting career paths and professional goals, you'll have a starting point for more in-depth research.

Remember, at any point, you can ask the AI to dig deeper into any intriguing observations it offers.

USING AI TO COMPARE CAREER PATHS

What if you're not sure about your career path? You can get insightful answers once you've fed accurate data into ChatGPT. One of my students, who was debating whether he should pursue a career in sales or project management, received the following response to this prompt:

→ **PROMPT:** *Based on this résumé, please give a compatibility rating as a % to him pursuing sales with a 2-paragraph description and then repeat for project management.*

ChatGPT RESPONSE:

Sales Role Compatibility: 80%

His extensive experience in driving business growth, executing marketing strategies, and managing key accounts demonstrate strong competencies for a sales role. His successful track record of strategizing, developing, and managing solutions that have addressed complex business challenges indicate a keen understanding of client needs and market trends. His skills in selling 6-figure training programs and taking a wellness business to seven figures in revenue clearly suggest his ability to close high-value deals. However, while his experience is impressive, the primary focus of his career has not been exclusively on direct sales, which may require additional training or transition time.

Project Management Role Compatibility: 90%

He has showcased strong project management skills throughout his career. His experience in leading cross-departmental initiatives, overseeing teams, managing vendor contracts, and successfully implementing complex projects aligns strongly with the expectations of a project management role. His leadership role in co-founding and managing his startup demonstrates his ability to steer a project from conception to completion. The fact that he led multiple enterprise innovation initiatives that were utilized by top companies such as National Geographic, Atlantic Media, and AARP further emphasizes his project management skills. His extensive experience in this area gives him high compatibility for a project management role.

➜ **PROMPT:** *Are there any other professional paths where the compatibility is above 90%?*

ChatGPT RESPONSE:

Innovation Consulting: Compatibility 95%

His résumé exhibits a high level of expertise in the field of innovation consulting. His diverse experience in strategizing, developing, and implementing innovative solutions to complex business challenges, as well as leading enterprise innovation initiatives, would make him an excellent candidate for senior roles in this field. As a cofounder and managing director, he has successfully created and implemented cross-departmental leadership development and mental wellness initiatives that surpassed financial goals. His understanding of emerging market trends, consumer insights, and technologies would be highly valuable in this role.

Leadership Development/Executive Coaching: Compatibility 90%

Another area where he might excel is in Leadership Development or Executive Coaching. His experience in developing innovative programming to guide and mentor executives at companies like Google, Microsoft, Salesforce, and Adobe, shows his aptitude for coaching and guiding top-level professionals. The integration of values such as care, respect, optimism, and resilience into leadership and staff training programs demonstrates his deep understanding of organizational behavior and leadership principles. His success in this area could translate into a very effective career as an executive coach or leadership development consultant.

****

CRAFTING YOUR RÉSUMÉ WITH AI

This section will be helpful if you are crafting your résumé for the first time or are building on an older version. The main sections of a résumé consist of:

- Accomplishments
- Skills
- Jobs
- Education

While many formatting and presentation philosophies exist, I will share some simple techniques you can use to effectively update your résumé, bringing it in line with current standards.

Step 1 – Ask your university career counselor to share résumé samples that can be helpful for you.

Step 2 – Use AI as illustrated in the following examples to update the content of your résumé *before* speaking with the career resource:

Materials Needed:

- Most updated résumé.
- Completed assessment content (related to career goals and accomplishments).

- Details of jobs held since the most recent résumé update.

> IMPORTANT: If your résumé is not optimized to be read by ATS applications (Applicant Tracking System), it may be rejected on formatting only. Advances in AI technology on the employer side also make it imperative you are aware of employers' formatting and content expectations.

Résumé Section: Headline

Headline Section: This is a brief introduction to your professional profile. It's a short statement or a title that appears at the top of your résumé after your name and contact details. The purpose of a headline is to quickly grab the recruiter's attention by summarizing your professional identity in a concise, compelling manner. It should highlight your most relevant skills, experiences, or roles. This area typically has your preferred job title and 4–6 bullet points showcasing the main attributes you offer a company. It starts with the job title you aspire to (which should be similar or exact to the job title you are applying for).

How do you get from having an outdated résumé or no résumé to having bullet points for your headline?

→ **PROMPT:** *You are a top-notch résumé coach and must update the following résumé. Make a summary of the main headline based on the following information: [paste résumé].*

➔ **PROMPT:** *Now take the résumé and make the bullets geared toward a _____ position.*

Résumé Section: Core Proficiencies / Skills

This résumé section focuses on keywords you can use to pass through ATS keyword filters. The words you use should reflect both current skill sets and the skills you are looking to use in your future job. AI can help you determine the majority of terms needed.

Here is an example of Ben's Core Proficiency section prior to starting his company:

CORE PROFICIENCIES

Artificial Intelligence | Machine Learning | Sales Enablement | Pre-Sales | Cross Functional Collaboration | NLP | New Business | Current Client Upsell | Competitive Selling | Global Sales | Team Leadership | (SaaS) | Integrations and APIs | Sales Demonstrations | Information Security | Innovation | Project Management | Quality Analysis | Coaching/Mentoring

---

Taking the work you have already completed, use the following prompt:

➔ **PROMPT:** *Under the core proficiency section of the résumé, please list the top 20 skills that summarize his abilities and experience.*

> IMPORTANT: Regardless of your major, the industry you are applying for or the role you want to have, make sure you add familiarity with generative AI. Examples of keywords: AI, Artificial Intelligence, ChatGPT, Generative AI. If you use DALL-E, Canva, or other AI tools, mention those.

Résumé Section: Accomplishments

The accomplishment section of a résumé focuses on tangible achievements throughout your career. For students that are graduating without work experience, you can leave this section out and put your achievements in the headline section.

If you have had the chance to gain work experience, here are some examples of how you populate this section:

➔ **PROMPT:** *Please create a list of accomplishments in the format of the accomplishments section of the résumé.*

Here is an example for one of my students:

ChatGPT RESPONSE

- Efficient Data Migration Implementation: Successfully managed the data migration of 250 legacy American Airlines planes into a single maintenance system, exceeding the targeted completion date by more than six months.

- **Cost-Saving AV Setup Process:** Developed an innovative approach to utilize surplus computer equipment for AV setup instead of purchasing new equipment, resulting in significant cost savings of $20,000.
- **User Error Messaging Project:** Spearheaded a project to improve user error messaging, which helped expand the company's sales by 20%.
- **Compliance Process Creation:** Designed and introduced a new compliance process for new team members covering policies, procedures, and regulatory requirements within the Technical Operations department.
- **Onboarding Process Improvement:** Overhauled the onboarding process for new hires, reducing the time to gain access to all documentation by 25%.

****

Résumé Section: Professional Experience/Other Sections

This section is the chronological description of your job title, duties, and accomplishments. I suggest taking what you have for each position and asking the AI to restate what you wrote. There are many times when GPT4.0 can offer some more powerful or succinct words that can improve the résumé.

Here is an example:

➔ **PROMPT:** *Improve this job description in the Experience section:*

American Airlines, Ft. Worth, TX
Sr Project Manager

- *Identified and analyzed project plans and timelines for projected deliverables and outcomes to shorten target completion date a year for data migration of 250 legacy American Airlines planes into a single maintenance system*
- *Spearheaded a newly implemented system of improved scheduling and access for all Records teams resulting in exceeding our targeted completion date by more than six months*
- *Implemented User Acceptance Testing for the benefit of the DevOps team, including new tracking of Supply Chain CR escalation process*

ChatGPT RESPONSE:

American Airlines, Ft. Worth, TX

Sr. Project Manager

- Devised and evaluated project plans, identifying strategic opportunities to expedite deliverables, which notably accelerated the completion timeline of a data migration project by a year. This project involved the consolidation of data from 565 legacy American Airlines planes into a single maintenance system.
- Championed the implementation of a new system that enhanced scheduling and access for the

Records team, significantly surpassing the initial target completion date by over 6 months.

- Directed the execution of User Acceptance Testing protocols, tailored to aid the DevOps team. This also included the development of a new tracking system for Supply Chain CR escalation processes, contributing to increased efficiency in operations.

****

On other résumé sections, such as Education and Languages, these data points are straightforward and do not need AI help.

BEYOND THE RÉSUMÉ: NEXT STEPS

There are many phases to having an effective job search including:

- Creating and updating your LinkedIn profile
- Find good job matches
- Applying to jobs and writing cover letters
- Interviewing
- Salary negotiations

To understand how best to leverage AI for these tasks, follow this QR code and order the book *Find Your Next Job with ChatGPT: A Guide for Everyone.*

SCAN ME

Beyond the Résumé Example 1 - Understanding Job Compatibility

When you begin targeted outreach; you have the necessary tools to evaluate whether an opportunity is a good fit, and you can optimize your chances.

While a job seeker can look at a job description and other demographic elements related to the company and opportunity to quickly determine if a job is in the ballpark of being a good fit, it would take 10–15 minutes to dig deep in the job description and see how good of a fit the job might be.

ChatGPT can give you a compatibility rating quickly and summarize the important points. Here is an example from Ben Gold:

→ **PROMPT:** *I am applying for a job. Please review the résumé and job description and give me an overall compatibility assessment as a %. Then give me a 2-paragraph description. After that, give me an evaluation of my top 3 strengths and top 3 weaknesses. Here is my résumé: [paste* *résumé]. Here is the job description: [paste job description].*

ChatGPT RESPONSE:

Overall Compatibility Assessment: 90%

Based on the provided résumé and job description, your qualifications seem highly compatible with the job requirements for the Sales Engineer position at ABC Technology. Your background in software pre-sales with a focus on AI and Machine Learning fits well with the company's emphasis on disruptive technologies in supply chain logistics. Your extensive experience across various industries like Healthcare, Finance, and Energy not only demonstrates versatility but also offers the opportunity for cross-industry applications of the technology you'll be working with.

Your technical proficiencies in AI, Machine Learning, APIs, and integrations are clearly strengths that should make you a strong candidate for the Sales Engineer role. The fact that you're already familiar with Linux, Python, and cloud-based software positions you well for a smoother transition. You also possess robust pre-sales and cross-functional collaboration skills, evidenced by your ability to work seamlessly with teams ranging from data science to sales and marketing. These are critical attributes for the consultative, problem-solving approach emphasized in the job description. Your only potential weakness might be the specific experience with Kubernetes and Docker, which are mentioned in the job requirements.

Top 3 Strengths:

1. **AI and Machine Learning Expertise:** Your strong background in software pre-sales focusing on AI and Machine Learning gives you an edge, especially considering the job's emphasis on these technologies.

2. **Cross-Functional Collaboration:** Your ability to work with different departments like sales, marketing, product, and data science teams makes you highly versatile and aligns perfectly with the job's requirement for a consultative, problem-solving approach.

3. **Technical Proficiencies:** Your understanding of APIs, Linux, and cloud-based software closely aligns with the job's technical requirements, setting you up for potentially seamless integration into the new role.

Top 3 Weaknesses:

1. **Kubernetes and Docker:** The job description mentions experience with Kubernetes and Docker as a preference. These specific technologies aren't noted in your résumé, so you may need to demonstrate competence in these areas.

2. **Industry-Specific Focus:** The new role is specifically in supply chain logistics for Fortune 500 companies. While you have broad industry experience, a deeper focus on this industry would strengthen your application.

3. **Depth of Programming Skills:** The job requires lite programming skills in Python, Java, Shell Scripting, or other languages. Your résumé doesn't specify the depth of your programming skills, which might require clarification.

****

More in-depth analysis available in the book by Ben Gold: *Find Your Next Job with ChatGPT: A Guide for Everyone.*

Beyond the Resume Example 2 - Crushing the Interview

ChatGPT can help a job seeker get personalized advice for job interviews. There are many types of interviews: recruiter screening calls, hiring manager interviews, executive-level interviews, and presentation expectations. The more data you can feed AI, the more personalized your preparation response will be.

The contextual data you will need is your résumé, the job description, and the LinkedIn profile of the person interviewing you. While the free versions of GPT can give you reasonable responses, this part requires 4.0.

Below are examples of several prompts and responses received in preparation for an upcoming interview.

➔ **PROMPT:** *I have an upcoming interview. Based on my résumé, the job description and the* *interviewer profile, share the top 5 items we can build rapport with because we share similar skills, backgrounds or interests. Here is my résumé: [paste résumé]. Here is the job description: [paste job description]. Here is the interviewer's profile: [paste profile].*

ChatGPT RESPONSE:

Based on your résumé, the job description, and the interviewer's profile, here are the top 5 areas that could help in building rapport:

1. **Sales Experience:** Both you and Joe have extensive experience in sales roles. Discussing the challenges and rewards of working in sales, as well as sharing some of your strategies for lead generation, sales conversion, and client retention, could help establish a common ground. You could also discuss the various industries you've worked in and how that broad experience has enhanced your sales skills.

2. **Technical Proficiency:** Your résumé shows strong technical knowledge in AI, Machine Learning, and APIs. Mr. Smith has a background in industrial automation and computer software sales. Sharing insights about how technology can facilitate sales and improve customer experience can be a good conversation starter.

3. **Leadership:** You've shown leadership skills by leading collaborative efforts with cross-functional teams and mentoring sales reps. Similarly, Mr.

Smith has a strong military background, implying leadership experience. Discussing leadership styles, team management experiences, and methods for empowering teams could foster mutual understanding.

4. **Client Onboarding and Transition:** Your role in ensuring successful software adoption and implementation by coordinating efforts with the training team for client and employee training aligns with the role requirement of transitioning the client to on-boarding. Sharing your experiences and best practices in this area could be a good talking point.

5. **Global Experience:** You've had extensive global experience and have worked with diverse teams. While Mr. Smith's profile does not explicitly mention global experience, discussing the benefits and challenges of operating in an international context could be insightful. Your language skills may also be an interesting topic.

➔ **PROMPT**: *What are the top 5 questions I will likely be asked?*

ChatGPT RESPONSE:

1. **Experience with HCM Sales:** Given that the job is for a Major Accounts HCM Sales Executive, you will likely be asked about your previous experience selling Human Capital Management (HCM) products. This could be something like: "Can you

describe your experience selling HCM solutions to major accounts?"

2. **Lead Generation and Conversion:** Since the job description mentions self-generating leads, qualifying prospects, and closing sales, you might be asked: "Can you describe your process for lead generation, qualification, and conversion? Can you share an instance where you successfully turned a lead into a major client?"

3. **Collaboration and Teamwork:** The interviewer may want to understand how well you work with others in a sales environment. You might be asked: "Can you share an example of a successful sales deal where you collaborated with cross-functional teams to meet the client's needs?"

4. **Handling Client Transition:** As the job involves ensuring a successful transition from proposal to onboarding, a possible question could be: "Can you describe your process for transitioning a client from the proposal stage through to successful onboarding?"

5. **Personal Sales Accomplishments:** The interviewer will want to understand your personal sales accomplishments. A typical question might be: "Can you share your most significant accomplishment in sales? How did you achieve it, and what was the impact on your previous company?"

****

More in-depth analysis available in the book by Ben Gold: *Find Your Next Job with ChatGPT: A Guide for Everyone.*

USING AI TO NETWORK WITH ALUMNI

Networking is an important skill for uncovering the hidden job market, and the standard networking vehicle for most job seekers is LinkedIn. The right connection could also help your fraternity or sorority chapter to succeed by getting access to more alumni resources or expertise. Updating your LinkedIn profile ensures you'll have as many quality connections as possible, and the larger your network is, the better chance a person you are reaching out to is connected to someone you know.

As a student who is looking to enter the job world, it is important to first understand resources available to create the best chance of building your network and in uncovering hidden opportunities. Here are the 3 main groups you can use:

1.   Fraternity or Sorority alumni
2.   School Alumni
3.   People on LinkedIn that are in the industries you want to learn about

Fraternity and Sorority Alumni

As a current student, you usually can tap into the alumni groups on LinkedIn. You will want to contact your fraternity/ sorority headquarters and see how much data they are willing to share with current students. You can also search directly on LinkedIn using your organization name as a search term and find many alumni there. The goal is to get a list with alumni and their current role and company.

Remember that because the bond of being a member of an organization is so strong, you should look more broadly at their experience. The goal of making these connections is not necessarily finding a job right away but to gain insights into their experiences, add to your network, and potentially benefit from their network.

University Alumni

Most career centers can assist you in locating alumni from your school based on your major or career path and identify people you can network with, including your fraternity/sorority alumni from your chapter. You should try to join university alumni groups (although groups will have different rules about admitting current students).

People on LinkedIn

While the bond is less strong from a random stranger, if you identify a person who is doing something you are genuinely interested in learning more about, a well-crafted and personalized message should land you a 15–30-minute conversation. They are more likely to accept a Zoom meeting with you because you are a college student and you have nothing to sell them—you're just genuinely interested in their career path, their employer, or how they became so successful.

Potential Networking Goals

- **Informational Interviews:** To gain in-depth insights about a particular role, industry, or company from a professional in the field.
- **Learn about a Company:** To understand the culture, values, mission, and work environment of a potential employer.
- **Learn about an Industry:** To grasp trends, challenges, and opportunities within a specific industry.
- **Uncover Job Openings:** To discover job opportunities that may not be widely advertised.
- **Expand Professional Network:** To meet professionals in your field or industry who can provide advice, mentorship, or job leads.
- **Personal Branding:** To articulate and refine your professional narrative and career goals.
- **Career Guidance:** To gain advice on career paths, necessary skills, and advancement strategies from seasoned professionals.
- **Mentorship:** To find a mentor who can guide you through your job search and career development.
- **Gather Industry Jargon:** To understand specific industry language, which can be used in résumés, cover letters, and interviews.

Once you have decided why you want to network, there are many ways to begin outreach:

#1 Create a prioritized list based on networking goals. Examples include:

- I would like to get 5 informational interviews from people who are product managers to learn about the role.
- I want to understand what entry level options are for engineers coming out of college
- I want to work at Google and learn about the company from current and former employees.
- I want to switch from sales to project management and speak to people with similar career choices.

#2 Create a database from the 3 sources above that you want to leverage to identify and narrow potential target individuals or companies based on location, size, industry, and role.

#3 Use LinkedIn to determine how best to filter your results:

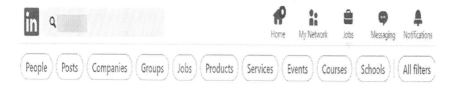

In the search bar, you can search by:

- People
- Posts
- Companies
- Groups
- Jobs

- Products
- Services
- Events
- Courses
- Schools

In addition to the search categories, you have in-depth filter options. For this networking exercise, I will drill into the filters you can use for People and Companies.

People Filter Options:

An important component is you can sort by connection levels:

1st = Direct connection

2nd = Connection of someone in common (in this case, you can evaluate who the common connection is and see if your common connection can make an introduction to the person you want to network with)

3rd = Connection distance between you and the person you want to network with.

Filter only People ▾ by   ✕

Connections

☐ 1st   ☐ 2nd

☐ 3rd+

Connections of

\+ Add a connection

Followers of

\+ Add a creator

Actively hiring (Premium)   Off ⬤

Locations

☐ United States   ☐ United Kingdom

☐ England, United Kingdom   ☐ New York City Metropolitan
                                      Area

☐ Israel   \+ Add a location

Talks about

☐ #startups   ☐ #leadership

☐ #innovation   ☐ #sustainability

☐ #technology   \+ Search a topic

Current company

☐ International Renewable Energy   ☐ Microsoft
   Agency (IRENA)

☐ Papaya Global   ☐ Salesforce Ben

☐ World Economic Forum   \+ Add a company

Filter only People ▾ by   ✕

School

☐ Ben-Gurion University of the Negev   ☐ University of Oxford

☐ Tel Aviv University   ☐ Bar-Ilan University

☐ University of Leeds   + Add a school

Industry

☐ Professional Services   ☐ Technology, Information and Media

☐ Technology, Information and Internet   ☐ Financial Services

☐ Manufacturing   + Add an industry

Profile language

☐ English   ☐ French

☐ German   ☐ Spanish

☐ Others

Open to

☐ Pro bono consulting and volunteering   ☐ Joining a nonprofit board

Service categories

☐ Consulting   ☐ Marketing

☐ Coaching & Mentoring   ☐ Operations

☐ Business Consulting   + Add a service

Filter only **Companies** ▼ by   ✕

Locations

☐ Europe   ☐ EMEA

☐ European Union   ☐ European Economic Area

☐ Schengen Area   **+ Add a location**

Industry

☐ Real Estate and Equipment Rental   ☐ Real Estate
  Services

☐ Medical Practices   ☐ Wholesale

☐ Hospitals and Health Care   **+ Add an industry**

Company size

☐ 1-10 employees   ☐ 11-50 employees

☐ 51-200 employees   ☐ 201-500 employees

☐ 501-1000 employees   ☐ 1001-5000 employees

☐ 5001-10,000 employees   ☐ 10,001+ employees

Job listings on LinkedIn

☐ Yes

Connections

☐ 1st

WAYS TO INTERACT WITH PEOPLE ON LINKEDIN

#1 Engage with a post from someone of interest.

Approaches to effective networking on LinkedIn have matured over time. A powerful strategy to engage and garner attention is showing genuine interest in someone's posts. Engaging with their content through likes or comments increases the likelihood of their reciprocation, whether by responding to a connection request or an InMail. The LinkedIn algorithm will also increase your chances of being found by professionals in your industry the more you comment, like, or engage with other users.

ChatGPT can play a pivotal role in this process by helping you craft thoughtful and intriguing comments. It assists in demonstrating genuine interest in their narratives, significantly improving your chances of kick-starting a 30-minute networking conversation.

→ **PROMPT:** *Based on the content of this LinkedIn post, write 3 short examples of insightful comments that would lead to a positive engagement with the poster [paste post].*

ChatGPT RESPONSE:

1. "Thanks for sharing these insightful statistics. It's clear that remaining cautious and grounded during the pre-seed stage is crucial for founders. Overvaluing too early can indeed lead to potential issues down the line. How do you recommend

founders balance this with the growing round size trends?"

2. "Interesting data points, particularly the fluctuating monthly valuations. It's fascinating to observe the MoM growth in average round sizes while maintaining reasonable valuations. I'd love to hear your thoughts on how this dynamic may evolve in Q3."

3. "Insightful post, particularly your advice for pre-seed founders. The balance between raising enough to fuel growth and avoiding overvaluation is indeed tricky. Could you expand on how these statistics may differ across various sectors?"

****

#2 Writing Connection Requests Where They Will Accept: 300-Character Limit

LinkedIn allows you to communicate with a person of interest when you attempt to make a new connection. When you have identified someone you want to network with, you can use the 300-character limit to expand your network and potentially get a conversation with the target audience.

First, download the PDF profile of the person you want to network with (located below their profile picture).

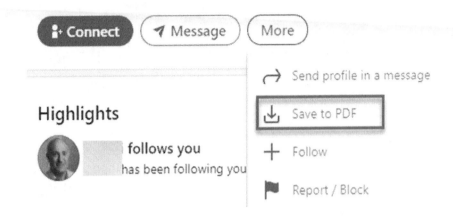

→ **PROMPT:** *You are a tenured career coach. I want to set up an informational interview to understand the best strategies to enter the digital marketing world. I am a senior and am a member of the same sorority as this LinkedIn member. First, describe to me the top 3 priorities of this person. Then give me 3 examples of a LinkedIn Connection request message that incorporates my understanding of their priorities in under 300 characters and will get their attention. Here is my résumé [paste résumé] and their profile: [paste profile].*

ChatGPT RESPONSE:

Top 3 Priorities of the Contact (Jane Smith):

1. **Driving Change and Building Brands:** Jane's career showcases a focus on transforming marketing strategies and building brand identities for various companies.

2. **Team Leadership and Coaching:** Jane appears to be heavily invested in leadership development, as evidenced by her role in coaching marketing executives.

3. **Results-Oriented Growth:** Jane places an emphasis on results that drive growth, including revenue generation and employee engagement.

3 LinkedIn Connection Request Examples:

1. Highlighting Shared Sorority & Interest in Digital Marketing

- Hi Jane, as a fellow SORORITY grad gearing up to enter the digital marketing world, I'm really impressed by your career in driving change and delivering results. Would love to learn from your insights during an informational interview. – Jessica

2. Focusing on Leadership & Coaching

- Hi Julie, I noticed we both are members of SORORITY and share a passion for leadership in marketing. I'd greatly value the opportunity to discuss how you coach marketing teams for success. Can we connect for an informational interview? – Jessica

3. Spotlighting Results-Driven Growth

- Hello Julie, I'm a SORORITY senior with a 3.95 GPA and a strong focus on results, much like your approach to growth in digital marketing. Would it be possible to schedule an informational interview to learn from your experiences? – Jessica

****

Other Prompt Ideas:

➔ **PROMPT:** *You are a tenured career coach. I want to send a message in the LinkedIn connection request box for an informational discussion about their background as a business owner. First, describe to me the top 3 priorities of this person. Then give me 3 examples of a LinkedIn Connection request message that incorporates my understanding of their priorities in under 300 characters. Here is their profile: [paste profile].*

➔ **PROMPT:** *This person is a LinkedIn influencer in an area I am interested in. Write me 3 examples of a compelling connection invite to show that I understand who they are. This invite aims to create a connection and get a 15-minute conversation.*

Copying the GPT text to the connection invite:

Below the profile picture, click the "More" button to see the
drop-down options. Click "Connect."

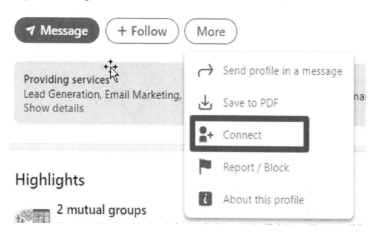

A new box opens up. Click the "Add a Note" button.

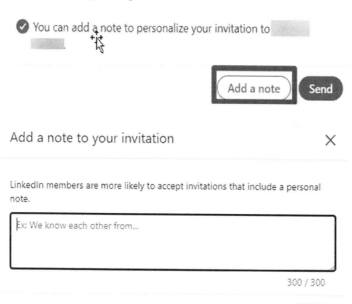

CLOSING THOUGHTS

We hope this has been helpful to expand your mind in terms of what's possible with Artificial Intelligence in the Fraternity and Sorority community. We believe that we have only scratched the surface, and we know that it will evolve quickly. Once you start using AI on a regular basis and practice with different prompts to see the different outcomes, there is no doubt that you will come up with new ideas and new applications of AI in your chapter and in your council. If you have new ways to use AI for your chapter or council that are not included in this book, we would encourage you to share them with us so we can make updates to this book and help all of Fraternity/Sorority Life to continue growing. Please send us an email with testimonials, updates, and suggestions to: bookings@greekuniversity.org.

If you are interested in speakers or consultants to engage with your Fraternity and Sorority community, please visit us at www.greekuniversity.org. You can also find us on social media @greekuniversity where we would love for you to follow along and interact with us.

To better understand how your fraternity or sorority experience will translate to success in the corporate world, follow this QR code and order the book *From Letters to Leaders: Leveraging Your Fraternity or Sorority Experience to Land Your Dream Job*.

ACKNOWLEDGMENTS

This book would not be possible without several people in the Fraternity and Sorority community, as we stand on the shoulders of giants. Thank you to Pete Smithhisler (Lambda Chi Alpha) for showing me the importance of our values and our ritual. Thank you to David Westol (Theta Chi) for showing me why we must fix the risk management problems within Fraternity and Sorority Life. Thank you to Dr. Paul Stumb (Sigma Chi), Dr. Bill McKee (Kappa Sigma), and Rusty Richardson (Phi Delta Theta) for reigniting my passion for higher education and allowing me to pursue my Master's Degree at Cumberland University. Thank you to Dr. Nicole Cronenwett (Alpha Omicron Pi) for being my mentor as I pursue a Doctorate degree at Middle Tennessee State University. Thank you to Dr. Kim Godwin (Delta Gamma) for serving as my Dissertation Chair and showing me the importance of qualitative research in the field of Fraternity and Sorority Life. You have all poured into me, and I will do everything I can to contribute back into our field for as long as I possibly can, and we will be sure to make it a better experience for the future undergraduate members in every organization and every council. I am forever grateful for your belief in my abilities. You have no idea how important that was (and continues to be) for me.

Finally, I want to thank Ben Gold, our Artificial Intelligence guru. He has learned an entirely new arena of higher education and has embraced us as his own. He understands the value of Fraternity and Sorority Life on college campuses, and has personally met with many undergraduates in focus groups to truly understand their needs and help, whether that be in their future careers or to recruit better members in their organization. We're very fortunate to have that expertise with us today, and I know that he'll be meeting all of you on your campuses to help you recruit more quality members than you've ever dreamed possible. Let's get to it, there's no time to waste. Recruitment will be here before we know it.

ABOUT THE AUTHORS

Michael Ayalon, M.S.P.

As a professional public speaker, author, social media figure, host of the Fraternity Foodie Podcast, and CEO of Greek University, Michael Ayalon is a recognized thought leader in the realms of fraternity and sorority life, business leadership, and higher education. He has headlined keynote presentations on over 200 college campuses in 40 states to help solve problems such as Sexual Assault, Hazing, Alcohol and Drug Abuse, and recruitment for college student organizations. Mike is able to take lessons learned from helping to build companies from startup to over $25 Million in annual sales, as well as best practices as the Former Executive Director of Sigma Pi Fraternity with 120 chapters and over 115,000 members, to create dynamic, positive, and results-driven keynotes and workshops that transform people's lives.

Mike and his team published the "From Letters to Leaders" book trilogy, which includes:

1. *Creating Impact on Your College Campus and Beyond*
2. *Redefining New Member Education and Leveraging Belonging to Eliminate Hazing*
3. *Leveraging Your Fraternity or Sorority Experience to Land Your Dream Job*

He is a graduate of the School of Management at the University at Buffalo, and has a Master's Degree from Cumberland University in Public Service Management. Mike is currently pursuing a Doctorate in Assessment, Learning, and Student Success (Higher Education Concentration) at Middle Tennessee State University and anticipates completion in 2025. Mike is also the recipient of the 2022 Ronald H. Jury Interfraternal Friendship Award from Phi Mu Delta National Fraternity.

To book Mike as a speaker on your college campus, visit www.greekuniversity.org/presentations

Ben Gold

Ben Gold is an AI Consultant, thought leader, and speaker who specializes in practical AI solutions for individuals and organizations. While most consultants stop at theory, Ben dives into actionable steps, working hands-on with clients to implement and optimize AI strategies and brings a fresh, real-world approach to leveraging AI for everyday success.

With two decades of experience in technology, Ben has honed his expertise in enterprise-level AI solutions for the past five years. Specifically, he has leveraged AI to analyze tens of thousands of customer conversations, providing invaluable insights for enterprise clients.

Ben's focus took a new direction in 2023 because of the explosive growth in generative AI tools. He took one of his passions of helping people defining their career paths and finding their dream jobs and combined it with his knowledge of generative AI to help hundreds of people succeed in their career goals. He engaged with recruiters, career counselors, and job seekers, and took his message to a wider audience through webinars, podcasts, personal coaching, and a recently published book *Find Your Next Job with ChatGPT: A Guide for Everyone*.

Ben has a global perspective that is shaped by a decade of professional experience in Europe, during which he learned to speak Romanian, Spanish, Italian, and German.

Ben also holds a distinctive accolade from the early 1980s. He was recognized as the first video game world champion, a feat immortalized in a *Life Magazine* photoshoot in November 1982, and winning a competition filmed on the TV show *That's Incredible*.

To book Ben as a speaker on your college campus, visit www. greekuniversity.org/ben-gold

Made in the USA
Columbia, SC
15 November 2023

26244494R00087